I0024153

The Low-Carbon Contradiction

CRITICAL ENVIRONMENTS: NATURE, SCIENCE, AND POLITICS

Edited by Julie Guthman and Rebecca Lave

The Critical Environments series publishes books that explore the political forms of life and the ecologies that emerge from histories of capitalism, militarism, racism, colonialism, and more.

The Low-Carbon Contradiction

ENERGY TRANSITION, GEOPOLITICS,
AND THE INFRASTRUCTURAL
STATE IN CUBA

Gustav Cederlöf

UNIVERSITY OF CALIFORNIA PRESS

University of California Press
Oakland, California

© 2023 by Gustav Cederlöf

Library of Congress Cataloging-in-Publication Data

Names: Cederlöf, Gustav, author.
Title: The low-carbon contradiction : energy transition, geopolitics, and
    the infrastructural state in Cuba / Gustav Cederlöf.
Other titles: Critical environments (Oakland, Calif.) ; 13.
Description: Oakland, California : University of California Press, [2023] |
    Series: Critical environments : nature, science, and politics ; 13 |
    Includes bibliographical references and index.
Identifiers: LCCN 2023013099 (print) | LCCN 2023013100 (ebook) |
    ISBN 9780520393127 (cloth) | ISBN 9780520393134 (paperback) |
    ISBN 9780520393141 (epub)
Subjects: LCSH: Energy conservation—Cuba. | Energy transition—
    Political aspects—Cuba. | Energy policy—Cuba. | Carbon dioxide
    mitigation—Economic aspects—Cuba. | Geopolitics—Cuba. |
    Infrastructure (Economics)—Cuba.
Classification: LCC TJ163.4.C9 C43 2023 (print) | LCC TJ163.4.C9 (ebook) |
    DDC 333.79097291—dc23/eng/20230405
LC record available at https://lccn.loc.gov/2023013099
LC ebook record available at https://lccn.loc.gov/2023013100

32   31   30   29   28   27   26   25   24   23
10   9   8   7   6   5   4   3   2   1

Contents

Illustrations and Table

TABLE

Preface

In the spring of 2007, a long decade marked by the lack of oil was just coming to an end in one of Latin America's most oil-dependent economies. My impression as I first arrived was that the reggaetón musician Daddy Yankee's megahit "Gasolina" was playing on repeat in every corner, a song about roaring engines and a hot-blooded desire for gasoline. But in Havana, public spaces were filled with messages of another kind. It was no different in western Pinar del Río or eastern Guantánamo. All over Cuba, billboards called for work efficiency and energy "saving" as a national, revolutionary undertaking: *"Ahorrando más, tendremos más. Una revolución con energía"* (Saving more, we will have more. A revolution with energy). I soon learned that I had arrived in the midst of the country's Energy Revolution. A project of radical environmental consequence, the Energy Revolution was decarbonizing the Cuban economy, but as I would later realize, it was also reconfiguring the socialist state in a more fundamental, political way.

This book builds on research I have carried out since my first encounter with Cuba, working over a year on the island and many more at a distance from it. My main research base in Cuba was the University of Pinar del Río "Hermanos Saíz Montes de Oca" (UPR) where a group of mechanical engineers in the Center for Studies in Energy and Sustainable Technologies

(CEETES) provided me with a workspace and tirelessly helped me sort out all necessary *trámites*—Cuban code for visas and research permits. I am especially grateful to Francisco Márquez Montesino for accommodating me in CEETES on repeated occasions and to Rolando Zanzi at KTH Royal Institute of Technology in Stockholm, who early on put me in contact with UPR. An institutional affiliation is a requirement for doing research in Cuba, and with the backing of UPR, I was able not only to speak to people informally but also to carry out interviews and make observations in state companies and to access research-only areas in archives. The institutional arrangement also gave me access to the university as a field site where I was able to take part in research and education and engage with scholars and students, many of whom were recruited from the industrial workforce in Pinar del Río.

Beside CEETES, my key entry points were the organizations CubaSolar and Fundación "Antonio Núñez Jiménez" de la Naturaleza y el Hombre (FANJ). I am particularly grateful to Bruno Henriquez at CubaSolar and Reinaldo Funes Monzote at FANJ for their generosity. I am also indebted to José Altshuler, Luis Bérriz, Mayra Casas Vilardell, Luis Guillermo Castillo González, and Francisco "Panchito" Lorenzo, who in different ways enabled my work and deepened my understanding of Cuban environmental politics.

I stayed over the long term in two households, one in Pinar del Río and one in Havana. Experiences and conversations with the members of these households have in large part informed my understanding of everyday energy use in the household setting. I learned from the most mundane tasks—making coffee while talking behind the veil of the *Radio Rebelde* broadcast—but also from more unusual events, such as discussing Engels's dialectics of nature during a storm-induced blackout. I leave the identities of my *familia pinareña* (Pinar del Río family) and my *familia habanera* (Havana family) undisclosed, but I am incredibly thankful for their kindness and support.[1] Coming to Cuba from a European background with institutional funding, I was reminded time and again of how my ability to access and use energy differed from that of my hosts, reflecting our different positions in the power geometries that shape energy use. The affiliation with UPR granted me residential status and access to spaces that non-Cubans normally were excluded from, but while I could travel from

Pinar del Río to Havana on the back of a hot, diesel-smelling truck for a nominal sum—just as any Cuban—I always had the ability to travel in an air-conditioned tourist bus too if I wanted. My argument that energy use, as a mundane socio-ecological practice, always raises questions of political economy is directly rooted in experiences like this.

Drawing on fieldwork in households and industry, the book is an attempt to examine socialist Cuba from the "inside." Much research on Cuba is still clouded by a Cold War logic in which bipolar conflict asserts narrative order, making Cuban history an extension of the US and Soviet empires. Based on conversations, observations, and documents from Cuba, my aim is instead to take Cuba's revolutionary history seriously and see it as a starting point for reconsidering the research priorities and conceptual frameworks that guide work on energy transitions and infrastructure in critical environmental studies today. The geographies of knowledge about energy are very particular, shaped as they are by research on a rather small set of countries in Europe and North America. An "inside" perspective on Cuban history, which still attends to processes and events that take place beyond national borders, provides an alternative outlook in social science energy research in an effort to, as Gavin Bridge writes, "theorize about energy geographies from elsewhere."[2]

To re- and deconstruct an official Cuban government narrative, I draw on archival material collected in the Biblioteca Nacional "José Martí," the Biblioteca Provincial Pinar del Río "Ramón González Coro," the library of FANJ, the library of CEETES, a government repository of political speeches, and the private collections of José Altshuler. I am grateful for the assistance of the staff in these libraries, and especially to Professor Altshuler for inviting me into his home. I have also benefited greatly from the newspaper collections in the Bodleian Library and the British Library and from the interlibrary lending service at the Maughan Library in the United Kingdom.

Cuban government and Communist Party archives remain closed for research. However, as Jennifer Lambe and Michael Bustamante argue, we should be careful not to overlook "the Revolution's own published archive" of newspapers, magazines, and bulletins as a source to the past.[3] During the Cold War, Western observers often drew a sharp line between the independent capitalist media and the tendentious socialist press—"Our press tries to contribute to the search for truth; the Soviet press tries to convey

pre-established Marxist-Leninist-Stalinist truth," as Fred Siebert and colleagues wrote in their media-studies classic *Four Theories of the Press*.[4] For
most communist parties, mass media were and still are regarded as active instruments in the production of the symbolic life of the Party and the
state, aiming not only to express opinion but to form it.[5] The dependence
of Cuban newspapers on state institutions leaves no doubt—and no pre-
tense to the contrary—that mass media are instruments of ideological production, just as political speeches and other government publications are.
The question, then, is not whether we can uncover a stable, impartial past
through these documents, but whose past they allow us to uncover. While
many Cubans in my experience find official representations to portray a
revolutionary hyperreality resonating poorly with their own lived experiences, we should ask how these representations have been co-productive of
lived experiences over time. As Lambe and Bustamante suggest, the Revolution's official narratives should be taken as an "analytical starting rather
than ending point," and that is my attitude in these pages.[6]

The book began life as a doctoral project at King's College London. I am
deeply thankful for the conversations I had with Raymond Bryant and
Alex Loftus during my time at King's. They provided detailed comments
and professional guidance in equal measures, and their transformative
work in political ecology has inspired my argument more than I think they
realize. Simon Batterbury, David Demeritt, Matthew Gandy, and Alf Hornborg also asked provocative questions on earlier drafts, and I can only
hope I have done justice to their feedback. In researching and writing,
I benefited from the insights and friendship of James Angel, Sophie Black-
burn, Corinna Burkhardt, Archie Davies, Cornelia Helmcke, Oscar Krüger,
Jon Phillips, and Alexandra Sexton. I received wisdom, too, from members of the Contested Development research group in King's College Geography, particularly Helen Adams, Christine Barnes, Andrew Brooks,
Ruth Craggs, and Richard Schofield. In the closing stages of writing, my
colleagues in the Department of Liberal Arts at King's and, latterly, the
School of Global Studies at the University of Gothenburg have been ex-
ceptionally supportive.

In its various stages, the research behind the book was funded by the
Graduate School at King's College London, the Economic and Social

Research Council, and the Royal Geographical Society (with IBG) in the United Kingdom as well as the Wenner-Gren Foundation in the United States. Parts of chapter 5 were previously published as "Maintaining Power: Decarbonisation and Recentralisation in Cuba's Energy Revolution." *Transactions of the Institute of British Geographers* 45(1): 81–94. I had the opportunity to present preliminary results and test various portions of the argument at the annual meetings of the American Association of Geographers, the Latin American Studies Association, the Political Ecology Network (POLLEN), and the Royal Geographical Society (with IBG). Two workshops were particularly useful: "Energy Infrastructure: Security, Environment and Social Conflict" held at Boğaziçi University in 2016 and "Energy, Culture and Society in the Global South" held at the University of Cambridge in 2019. I would especially like to thank Jennifer Baka, Gavin Bridge, Donald Kingsbury, Reinaldo Funes Monzote, Gordon Walker, and Paul Warde for collaboration, feedback, or simply useful exchanges at these events.

It goes without saying that the largest source of support is my family: Gunnel and Leif Cederlöf, Erik and Elin Täufer Cederlöf, Laila and Niels-Peter Hansen. Hiking with my parents among the *vegas* and *mogotes* in Viñales was one of the highlights during my lengthy stay in Pinar del Río in 2015. I want to say extra thanks to my mother, Gunnel, for discussions big and small, for encouragement and always listening to unfinished thoughts. Beyond comparison, though, it is Vanessa Hansen who carries me through life. She has walked with me along the Thames and sat with me on the Malecón. We now carry Frank together, and he has helped too, insisting on the importance of raspberries, swings, and watering the garden.

Gustav Cederlöf
Gothenburg
March 2023

Note: For readers less familiar with Cuba, it will be helpful to know that the Cuban currency is the peso (CUP). The peso is subdivided into centavos. In 1993, the US dollar was made legal tender, which established a dual economy. The Central Bank of Cuba then replaced the dollar with a new currency in 2004, the convertible peso (CUC). The CUC was pegged to the dollar and could be traded for 24 CUP. In a long-awaited monetary reform, the currency was reunified in 2021.

Acronyms

AFP	American and Foreign Power Company
ALBA	Alianza Bolivariana para los Pueblos de Nuestra América (Bolivarian Alliance for the Peoples of Our America)
ANIR	Asociación Nacional de Innovadores y Racionalizadores (National Association of Innovators and Rationalizers)
BTJ	Brigadas Técnicas Juveniles (Technical Youth Brigades)
CAI	Central Agroindustrial (Agroindustrial Central; sugar mill)
CCGT	combined cycle gas turbine
CDR	Comités de Defensa de la Revolución (Committees for the Defense of the Revolution)
CEAC	Comisión de Energía Atómica de Cuba (Cuban Commission for Atomic Energy)
CEETES	Centro de Estudios de Energía y Tecnologías Sostenibles (Center for Studies in Energy and Sustainable Technologies)

CEN	Central Electronuclear (Nuclear Power Plant)
CERN	European Organization for Nuclear Research
CITMA	Ministerio de Ciencia, Tecnología y Medio Ambiente (Ministry of Science, Technology, and Environment)
CMEA	Council for Mutual Economic Assistance (also known as Comecon)
CNE	Comisión Nacional de Energía (National Energy Commission)
CTC	Confederación de Trabajadores de Cuba (Confederation of Cuban Workers, 1936–61)
	Central de Trabajadores de Cuba (Cuban Workers' Central Union, 1961–)
CTE	Central Termoeléctrica (Thermoelectric Power Plant)
CUC	convertible peso
CUP	Cuban peso
ECLA	United Nations Economic Commission for Latin America
EJ	exajoule (10^{18} joules)
EPG	Empresa Pecuaria Genética (Genetic Livestock Company)
FANJ	Fundación "Antonio Núñez Jiménez" de la Naturaleza y el Hombre ("Antonio Núñez Jiménez" Foundation for Nature and Man)
FMC	Federación de Mujeres Cubanas (Federation of Cuban Women)
GDP	gross domestic product
GE	General Electric
GOELRO	State Commission for the Electrification of Russia
IAEA	International Atomic Energy Agency
INRA	Instituto Nacional de Reforma Agraria (National Institute of Agrarian Reform)
JUCEPLAN	Junta Central de Planificación (Central Planning Board)

LPG	liquefied petroleum gas
MINAG	Ministerio de la Agricultura (Ministry of Agriculture)
MINAZ	Ministerio del Azúcar (Ministry of Sugar)
MINBAS	Ministerio de la Industria Básica (Ministry of Heavy Industry)
MINCEX	Ministerio del Comercio Exterior y la Inversión Extranjera (Ministry of Foreign Trade and Foreign Investment)
MINCIN	Ministerio de Comercio Interior (Ministry of Interior Trade)
MINEM	Ministerio de Energía y Minas (Ministry of Energy and Mines)
MININD	Ministerio de Industrias (Ministry of Industries)
M-26-7	Movimiento 26 de Julio (26th of July Movement)
NASA	National Aeronautics and Space Administration (United States)
NEP	New Economic Policy (Soviet Union)
OECD	Organization of Economic Cooperation and Development
OPEC	Organization of the Petroleum Exporting Countries
PAEC	Programa de Ahorro de Electricidad en Cuba (Program of Electricity Saving in Cuba)
PCC	Partido Comunista de Cuba (Cuban Communist Party)
PDVSA	Petróleos de Venezuela, S.A.
PSP	Partido Socialista Popular (Popular Socialist Party)
SDI	Sustainable Development Index
SDPE	Sistema de Dirección y Planificación de la Economía (System of Economic Management and Planning)
SEN	Sistema Electroenergético Nacional (National Electricity System)
SSR	Soviet Socialist Republic
UNDP	United Nations Development Program

UNE	Unión Eléctrica
UPR	University of Pinar del Río "Hermanos Saíz Montes de Oca"
US(A)	United States (of America)
USSR	Union of Soviet Socialist Republics

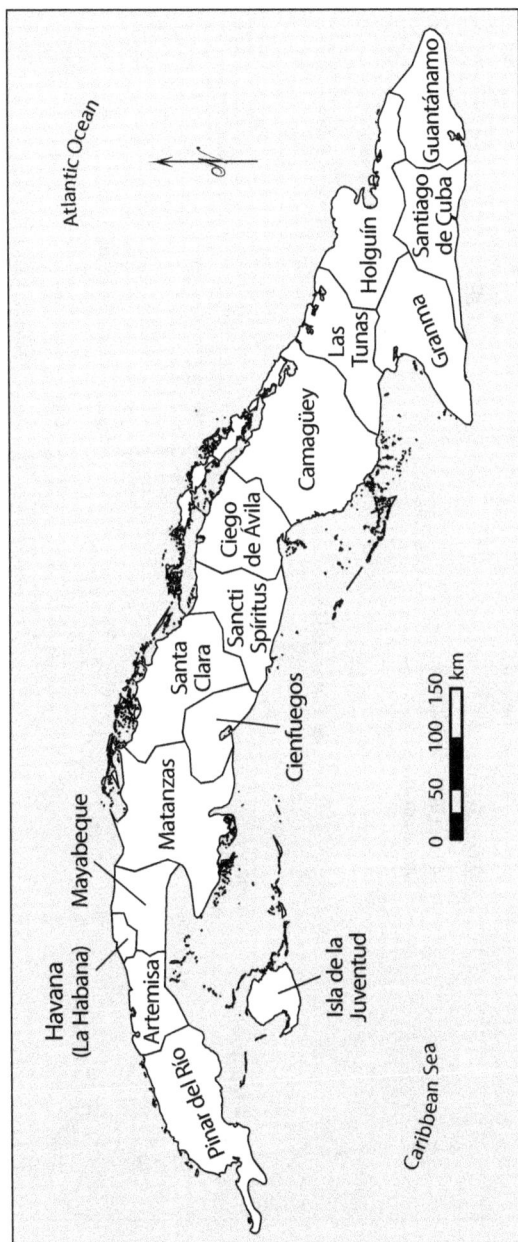

Map 1. Cuba. The provinces of Artemisa and Mayabeque were established in 2011, before which they together formed the rural province of La Habana. At that time, the capital city was known as Ciudad de La Habana. The current system of provincial government was created in 1976 during the so-called institutionalization of the Revolution.

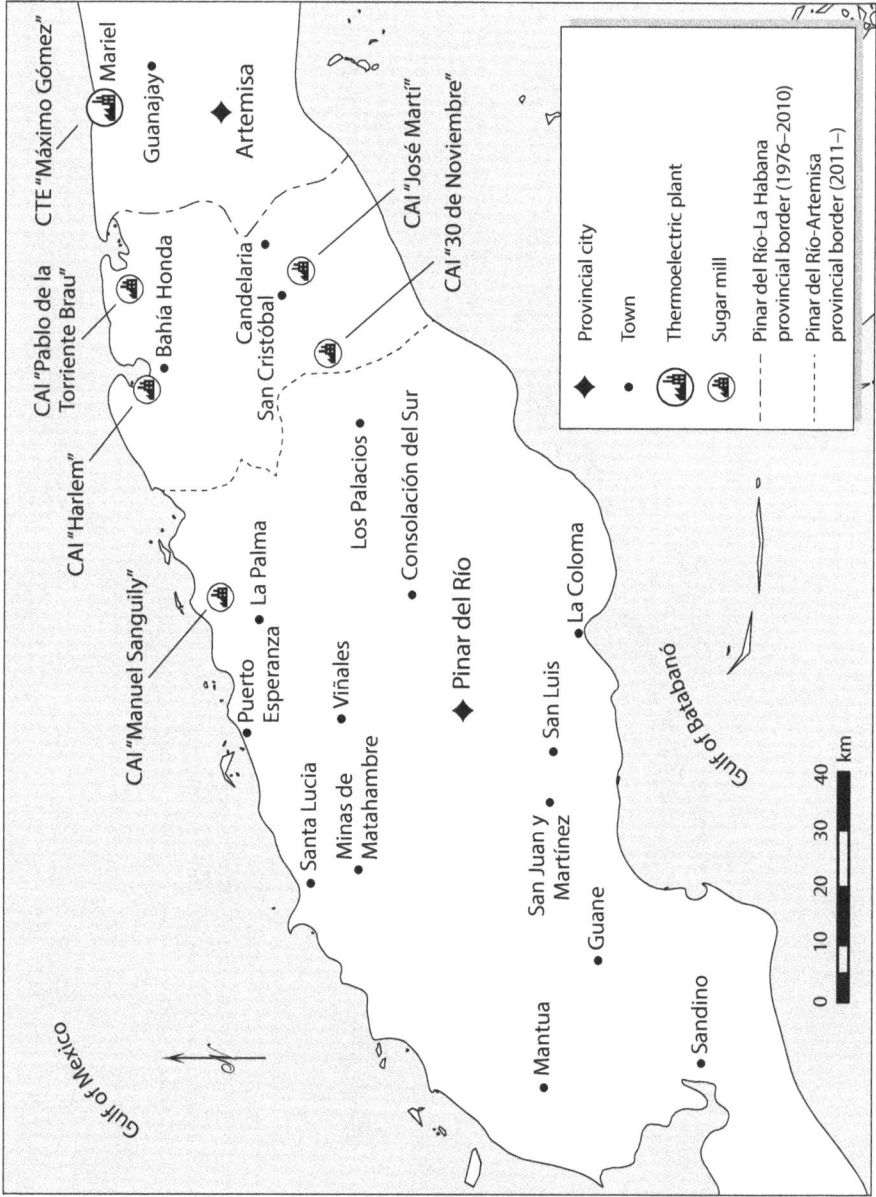

Map 2. The Province of Pinar del Río

Legend:

- ◆ Provincial city
- • Town
- 🏭 Thermoelectric plant
- 🏭 Sugar mill
- –·–·– Pinar del Río–La Habana provincial border (1976–2010)
- ········· Pinar del Río–Artemisa provincial border (2011–)

Labeled locations:

Gulf of Mexico

CTE "Máximo Gómez"
Mariel
Guanajay
Artemisa
CAI "Pablo de la Torriente Brau"
Bahía Honda
CAI "Harlem"
CAI "José Martí"
Candelaria
San Cristóbal
CAI "30 de Noviembre"
CAI "Manuel Sanguily"
La Palma
Puerto Esperanza
Viñales
Los Palacios
Consolación del Sur
Santa Lucía
Minas de Matahambre
Pinar del Río
San Luis
La Coloma
San Juan y Martínez
Guane
Mantua
Sandino
Gulf of Batabanó

0 10 20 30 40 km

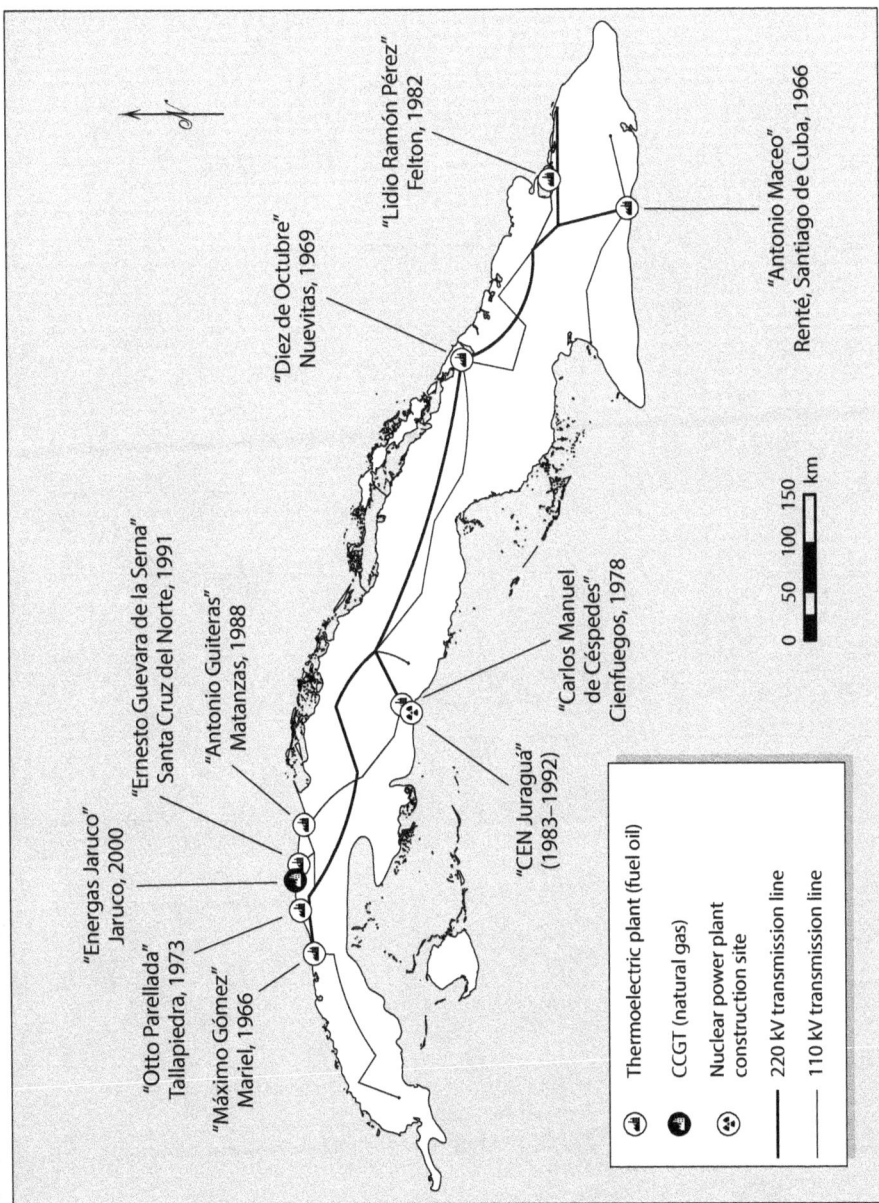

Map 3. The National Electricity System (SEN) with large power plants

"Energas Jaruco"
Jaruco, 2000

"Ernesto Guevara de la Serna"
Santa Cruz del Norte, 1991

"Antonio Guiteras"
Matanzas, 1988

"Otto Parellada"
Tallapiedra, 1973

"Máximo Gómez"
Mariel, 1966

"Diez de Octubre"
Nuevitas, 1969

"Lidio Ramón Pérez"
Felton, 1982

"Carlos Manuel
de Céspedes"
Cienfuegos, 1978

"CEN Juraguá"
(1983–1992)

"Antonio Maceo"
Renté, Santiago de Cuba, 1966

0   50   100   150
km

- Thermoelectric plant (fuel oil)
- CCGT (natural gas)
- Nuclear power plant
  construction site
— 220 kV transmission line
— 110 kV transmission line

Introduction

Reinier's refrigerator, a Soviet manufactured "Minsk," was not only empty but also hopelessly warm. While food was difficult to come by in the first place, there was little point in storing anything in it because of the power cuts. On Monday June 7, 1993, the blackout started at seven in the evening. According to the timetable that had been published in the provincial newspaper *Guerrillero* the week before, the state utility had shut off the electricity supply to Pinar del Río's third district already once that day. The current blackout was scheduled for five hours in three of the provincial distribution networks.[1] The city came to a halt, energyless, but for Reinier and his family it was nothing out of the ordinary. Just as food was in short supply, blackouts occurred almost every day of the week, every week, throughout the 1990s. The problem was greater than a blown fuse: it was the collapse of the Soviet Union.

Four decades earlier, Fidel Castro had outlined his visions for a new independent Cuba, promising that once a progressive government was in power, electricity would reach "to the last corner of the Island."[2] The revolutionary program of land reform, alphabetization, and improved public health was undergirded by a vision of energy use. Heading the Ministry of Industries, Ernesto Che Guevara argued that electrification was necessary

1

for Cuba's transition to communism. Electricity infrastructure constituted a techno-material base that enabled industrialization and automation. Nationally integrated energy infrastructure overbridged the development gap between the city and the countryside, the wealthy entertainment districts and the poor *barrios*. By the mid-1980s, a national electricity system—the SEN—interconnected Cuba from Pinar del Río to Guantánamo via Havana, Santa Clara, and Camagüey. In the revolutionary narrative, the infrastructure did historical work: it enabled the modernization of Cuba, reduced social difference, and, on these grounds, induced communism. The Revolution, one might say, had infrastructural form.

A series of thermoelectric power plants powered up the SEN, and they were named after heroes of the Cuban anti-colonial struggle. In a thermoelectric plant, an energy-potent material—often a fossil fuel—is put on fire. Some of the resulting heat is used to turn water into steam. The steam is led into a turbine, which is connected to a generator that converts mechanical energy into electricity. The state utility Unión Eléctrica made use of fuel oil to set this chain of events in motion. Following a first trade deal in 1960, the revolutionary government imported fuel oil from the Soviet Union in exchange for sugar. As Figure 1 shows, Cuba imported 13.3 million tons (Mt) of oil in 1989, almost all of which originated from oilfields in West Siberia. In the 1980s, the socialist state also launched a nuclear program aiming to replace the oil-fired energy system with one based on nuclear energy. Both Cuban and Soviet leaders spoke of their exchanges as expressions of socialist fraternity and fair trade: "The USSR has given our people terms of commercial exchange and long-term credits that constitute a true model for relations between a large industrial country and a small nation," Fidel Castro declared to a mass crowd during the first visit of a Soviet head of state to Cuba. "One million Cuban patriots express . . . their indestructible friendship, deep affection and eternal gratitude to the USSR," he said before the sea of people broke into chants of "Brezhnev, Brezhnev!" and "United Cuba and the USSR will win!"[3] Cuba's power plants, the oil they combusted, and the oil's terms of exchange were part of an imaginary of national liberation and socialist development, and the infrastructural system operated not only technically but also in the political economic and the semiotic domains.[4]

As Figure 1 also shows, the situation changed dramatically in 1990. Between 1989 and 1995, Cuba's imports of crude oil decreased by 86 percent

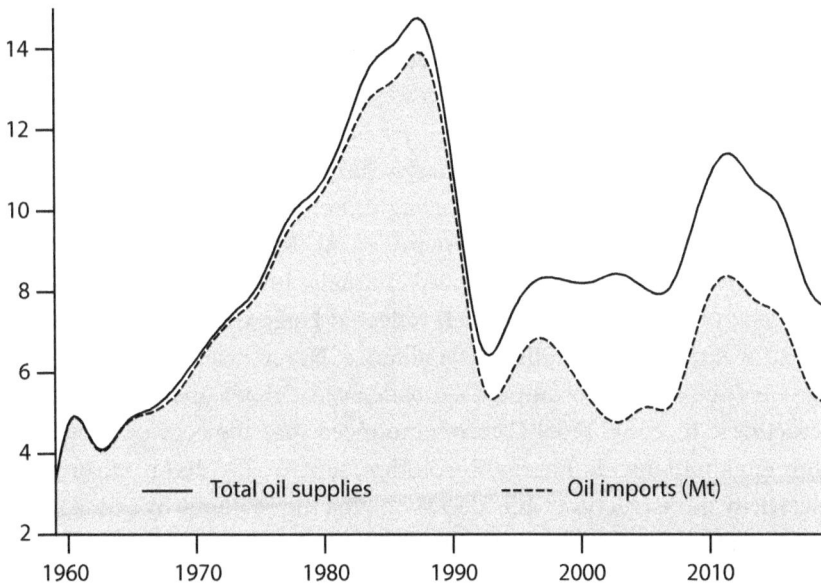

Figure 1. Cuban supplies of crude oil and oil derivatives, 1959–2019. *Sources:* ONE, *Estadísticas energéticas en la Revolución*, tables 14, 25; ONEI, *Anuario estadístico de Cuba 2014*, tables 10.4, 10.7; ONEI, *Anuario estadístico de Cuba 2019*, tables 10.3, 10.6; ONEI, *Anuario estadístico de Cuba 2020*, tables 10.3, 10.6.

and the availability of oil derivatives—fuel oil, diesel, kerosene, gasoline, and liquefied petroleum gas (LPG)—declined by 47 percent (1989–93).[5] There were four geopolitical reasons behind the energy crisis. First, Cuba lost its beneficial oil trading agreement as a result of the Soviet collapse. This loss coincided, second, with spiking international oil prices following Iraq's invasion of Kuwait; third, with a slump in the sugar market; and fourth, with the US government reinforcing its economic blockade of the island. In workplaces and households, the effects were direct and far reaching. After more than a century of sugar monoculture and decades of mechanized farming, the island's cane fields were lifeless without synthetic inputs. Tractors and trucks were immobile without diesel and gasoline. The state's reserves of kerosene and LPG, rationed for cooking purposes, ran out in 1993. In the SEN, the electricity supply was at best intermittent. While Cubans were forced to adapt to a new low-carbon reality, Fidel Castro announced that the country had entered a "special period in times of peace."[6]

The situation slowly improved before disaster struck again. This time the shock was climate related rather than geopolitical. In August 2004, hurricane Charley wreaked havoc in western Cuba; "never before had the province been in such conditions," *Guerrillero* summarized the events once the worst was over.[7] Hurricane Charley disconnected the entire province of Pinar del Río from the SEN, leaving more than six hundred thousand people without electricity for eleven days. At this time, new oil supplies were arriving in Cuban ports from Venezuela, in large part in exchange for Cuban medical services, which reflected Hugo Chávez's bid to create a Caribbean anti-imperialist trade alliance. Nevertheless, the collapse of the electricity system prompted a drastic overhaul of Cuba's energy infrastructures. In 2005, Fidel Castro announced that the country would go through a nationwide Energy Revolution, territorially decentralizing the electricity infrastructure while decarbonizing the economy by over a third. In reaching this goal, the Energy Revolution also did something more: it fundamentally changed the political nature of the socialist state.

THE LOW-CARBON CONTRADICTION

Cuba's post-Soviet experience figures vividly but often anecdotally in discussions on energy transitions. In *Societies beyond Oil*, John Urry refers to the "Cuban miracle," asserting that Cuba has a life expectancy on a par with the United States but "uses only about one-tenth of the USA's energy per person."[8] In 2006, the Worldwide Fund for Nature reported that Cuba was the only country in the world to combine a high Human Development Index with an ecological footprint kept within the limits of the biosphere, thus making it the only country to have achieved "sustainable development."[9] When a new Sustainable Development Index (SDI) was developed in 2015, Cuba was again ranked first in global comparison.[10] Many authors frame Cuba's special period in a narrative of simulated "peak oil"—the point in time when the availability of oil enters a terminal decline. Running with this metaphor, Cubans are said to have encountered an "abrupt and imposed" oil peak in the 1990s, after which they developed a low-carbon economy on the basis of economic and social reform rather than high-tech innovation. While peak oil represents a point in time when oil production reaches its maximum level—an apex rather than a historical low—the special period can provide an approximate understanding of the effects of declining oil supplies at the local and national levels.[11] Based on these accounts, then, Cuba's post-Soviet history appears to offer a model for a radical low-carbon transition.

The agricultural sector has been the predominant focus for studies of Cuban low-carbon development. The revolutionary government imported food from the socialist bloc during the Cold War, freeing up space for sugar production, but the geopolitical upheavals in the early 1990s forced it to cut back on its rationing of food.[12] It was difficult to increase the national levels of food production since the long-standing use of synthetic pesticides and fertilizers made it hard to farm the island's impoverished soils without them. While Cubans lost weight, a neuropathy epidemic—a neurological disorder affecting the eyes—spread in the country, likely resulting from vitamin-B deficiency.[13] In response, many farmers began employing agroecological methods in rural areas, to farm without agrichemical inputs, while a popular movement of organic urban agriculture emerged in cities. In 2021, the government's plan target was for 10 m^2 per inhabitant to be cultivated in urban and suburban farms.[14] This process of agrarian change, which according to one sympathetic observer was "the largest conversion from conventional agriculture to organic and semi-organic farming that the world has ever known," went hand in hand with conversations on food sovereignty and the development of horizontal knowledge sharing networks among small farmers.[15]

Cuba's post-Soviet experience is also mooted by advocates of degrowth. The call for degrowth starts with the observation that exponential economic growth ultimately is impossible to sustain on a planet with a limited biosphere. As Vaclav Smil shows, the historical data are unequivocal: up to the present, the levels of economic growth, energy consumption, and carbon emissions have followed the same incremental curves almost exactly. In order to transition to a low-carbon economy at a time when the global mean temperature is already increasing fast, Smil argues that this close coupling "makes it highly misleading to advocate any growth-oriented policies [even] assuming that . . . decoupling, and continued GDP growth, is possible."[16] For Giorgos Kallis and colleagues, it is fossil fuels that have made a history of growth possible. If energy is a "source of useful work," fossil fuels have enabled growth by doing "things human labor alone could not do." A low-carbon economy is therefore likely to be one of reduced productivity and output, seeing that renewables are more diffuse in space than fossil fuels are.[17] The argument for degrowth is usually pitted against capitalism, the very foundation of which is capital accumulation, but it is necessarily also an argument against the socialism that developed in Cuba, the Soviet bloc, and China in the twentieth century. The state-socialist aim was to spur rapid industrialization and to outcompete the capitalist West in the pursuit of "development."[18]

The special period adds further impetus to the degrowth critique: without a continuous supply of oil, Cubans were unable to sustain the socialist economy founded on growth. While most mainstream approaches to sustainable development treat the biosphere and the economy as complementary and substitutable dimensions of human activity, the economy may therefore better be conceptualized as a subsystem of the biosphere. Discussing the SDI, Jason Hickel details that within this model a country's level of income is so strongly coupled with a large ecological footprint that high income precludes the country from being ecologically sustainable. The SDI therefore incorporates a threshold so that countries with relatively low levels of income, such as Cuba, score more highly, all other things being equal. If Hickel's observation is correct, the countries with the highest levels of income could only become sustainable if they significantly de-grew their economies, reducing the throughput of energy and raw materials in production and consumption.[19]

The degrowth goal "is to build a society in which we can live better whilst working less and consuming less."[20] Such a project requires radical material changes to occur in economic processes at all scales, and particularly in the parts of the globalized economy with the highest rates of consumption. However, a smaller metabolism is only one aspect of successful degrowth. Seeing that growth is the backbone of the Western development paradigm, in both its capitalist and socialist iterations, a community pursuing degrowth would have to redefine the primary drivers of the economic process to instead base it on a cultural ideal of a "good life" other than growth-based modernity. As Federico Demaria and Ashish Kothari write, "In a degrowth society everything will be different from the current mainstream: activities, forms and uses of energy, relations, gender roles, allocations of time between paid and non-paid work, and relations with the non-human world."[21]

The degrowth critique sustains a narrative in which absolute planetary boundaries, the stability of the global climate system, and the laws of thermodynamics variously are emphasized as the limiting factors to exponential growth. On the one hand, the reading of Cuba's post-Soviet history in terms of a simulated peak oil scenario ties into this narrative, reflecting the argument that Cubans have adapted to a set of externally imposed biophysical limitations. On the other, the narrative echoes Malthusian arguments of old, such as those set out in the Club of Rome's report *Limits to Growth*, which suggest that humans either must adapt to absolute biophysical limits or else succumb to them.[22] However, more than insist on the existence of absolute limits, the degrowth proposal calls on communities to impose limits on themselves. Instead of reacting to an externally imposed limit, a community empowers itself to develop new forms of socio-ecological organization through a process of voluntary self-limitation.[23] The Cuban case is particularly interesting in this regard, because the biophysical limits to growth were not absolute but socially produced—they were geopolitically imposed—and possibly Cuba's special period can offer important insights on a degrowth future.[24]

It is important in this context to recognize that degrowth is not tantamount to economic recession; rather, it is an endeavor to organize economic life beyond the growth imperative, allowing production and consumption to take place under qualitatively different conditions.[25] In this sense, Cuba's post-Soviet history is an imperfect example of degrowth. While many Cubans built toward low-impact livelihoods, the special period was far from a voluntary political project, and the growth imperative has remained at the core of the Cuban Communist Party's (PCC) economic program. Thus, while many aspects of Cuba's special period can be interpreted in a degrowth framework, it is often closer to hand from the perspective of Cuban government and Party policy to argue that Cuba offers a window on efforts to develop an eco-socialist economy. As opposed to degrowth, this is a political vision in which growth and large-scale infrastructure are seen as the requisite conditions for economic development and social equality. Growth creates an abundance that can be shared equitably among its producers. The history of revolutionary Cuba is replete with examples of

large infrastructural solutions to social, economic, and environmental issues, resonating with a modernist socialist and, more recently, eco-socialist
ideal. The construction of the SEN, Cuba's nuclear program, and the Energy Revolution, as we shall see, are three cases in point.

Thus, it is fair to say that strong tensions are at play between degrowth-
inspired and eco-socialist trajectories in recent Cuban history. At the core
of this book is also a Cuban contradiction of energy transitions. The contradiction is this: (1) life in revolutionary Cuba has been shaped by a political project embroiled with fossil-fueled economic growth, but (2) life
in revolutionary Cuba has been shaped by political efforts to create a low-
carbon economy resting on non-growth-based social values. There have
been comprehensive efforts to develop a non-capitalist economy resting
on specific modes of energy use—fossil, nuclear, and low-carbon—but these
efforts have all been incomplete, fractured, and forfeited. Instead of suc-
cumbing to the temptation of simplification, the Cuban contradiction
should be studied in its complexity: by exploring the tensions between
degrowth-inspired and eco-socialist ideals, paying attention to "actually
existing geographies and their politics," we gain a better understanding of
how energy transitions rework political, economic, and social relations in
the pursuit of a more sustainable future.[26]

FUTURES PAST: ENERGY TRANSITIONS
AND ENVIRONMENTAL HISTORY

The study of two events has overwhelmingly shaped our understanding of
energy transitions: the Industrial Revolution in the United Kingdom and
the urbanization of the United States. A common feature of this scholarship is the way it foregrounds energy resources as objects of study—usually
water, coal, and oil—and envelops them in narratives of capitalist development, increasing mobility, political struggle, and an unprecedented "great
acceleration" of human activity.[27] Moreover, these historical energy transitions tend not only to appear as empirical facts in the literature but also as
historiographical devices. The shift from water to coal in Britain, or from
coal to oil in the United States, plays an organizational role in the way narratives of historical change are strung together. The turn from one energy

source to another signals a qualitative change from a more basic to a more modern economic system, thus periodizing time in eras before and after. It is not unusual for histories of energy transitions to reproduce teleological views, evoking the "stadial" development theories so emblematic of nineteenth- and twentieth-century modernity, which explain coincident differences in space as differences in time. However, seeing that these development theories were often formulated in a context of increasing fossil fuel use—by people experiencing the positive effects of it—we need to take a critical stance toward them.[28]

Whether framed in developmentalist terms or not, it is evident that energy transitions are processes of extensive environmental transformation. In *The Path to Sustained Growth*, E. A. Wrigley focuses on the transition from water to steam power in eighteenth-century Britain, and he identifies a shift from an "organic" to a "mineral economy" in the transition. The organic economy, he argues, drew its low entropy needs horizontally from rivers, fields, and forests, and as a result, the metabolism of the industrial economy was constrained by the rate of insolation and the availability of land. The mineral economy, by contrast, tapped into vertical accumulations of decomposed organic material, drawing energy from subterranean stocks of fossil fuels. The ability to "burn buried sunshine," to use Jeffrey Dukes's memorable phrase, or to tap into the "subterranean forest," to use Rolf Peter Sieferle's, freed up large swathes of land for other productive uses.[29]

The British and US energy transitions thus point to a radical geographical reconfiguration implicit in the transition from renewable to fossil energy. By the 1890s, coal provided the British economy with energy that otherwise would have required timber from forests eight times the size of the country's land area.[30] The mechanization of US agriculture in the 1920s is estimated to have freed up one-quarter of the country's farmland for other uses, which had previously been used to grow fodder for horses.[31] If we compare the Cuban censuses of 1846 and 1862—a period of rapid expansion in the colonial sugar industry—the island's forested area was also recorded to have decreased by 60 percent. This is likely an exaggeration, but the figure speaks to the scale at which wood and charcoal were used as fuels in the sugar mills and forestlands were converted into cane fields for colonial extraction through a swidden agricultural system.[32]

Clearly, energy production requires space, but different forms of energy also contribute to the historical production *of* space. As organic and mineral economies make different demands on land to facilitate the metabolic reproduction of socioeconomic relations, they generate historically and geographically specific patterns of energy extraction, distribution, and consumption.[33] Thus, for Smil, any urban and industrial economy pursuing a low-carbon transition "will require a profound spatial restructuring of the existing energy system, a process with many major environmental and socioeconomic consequences."[34]

When insights on energy transitions gleaned from Anglo-American experiences are transferred to a postcolonial context, they need to be qualified in at least two ways. First, we should be careful about making a sharp analytical distinction between an organic and a mineral energy regime. Wrigley stops short of noticing the continuities that exist between organic and mineral modes of energy use, as indeed there have been few clean breaks between the use of two energy sources in history. As chapters 1 and 2 will show, the construction of an oil-fired electricity system in Cuba was of critical importance for the socialist revolution, but biomass continued to be used in large quantities as a fuel in the sugar industry throughout the twentieth century; a large part of the population also continued to cook with kerosene and LPG despite the "modern" connotations of electricity. When Cubans faced an acute lack of oil in the special period, as chapters 3 and 4 examine, they also embarked on a low-carbon transition, but in the face of oil scarcity, the socialist economy remained heavily oil dependent (the focus of which is chapter 5). In the globalized economy, oil of course remains the quintessential fuel, and yet the overall consumption of coal has never been higher than it is today.[35] The old and the new instead tend to be used in parallel, reflecting the utility of different energy forms and technologies for different economic and sociocultural purposes. A global history of energy transitions is therefore more accurately read as "a story of aggregate increase and intensification."[36]

Second, narratives of energy transition tend to overlook the place of Britain and the United States in the international political economy. Indeed, while the deforestation and changing land uses of the British Isles served the incipient industrial economy, so did the deforestation and changing land uses of the Caribbean island economies. While coal metabolized the cotton mills in England and Scotland, industrialization continuously drew on organic economies throughout the British Empire, relying on access to vast swathes of land at a distance from the British Isles.[37] Any concept of energy transition must be sensitive to how unequal power relations locally but also globally condition it in order not to remain historically and geographically one-eyed. The conditions of possibility for an energy transition to take place in a postcolonial setting, such as the Cuban, are unlikely to mirror those in the core of a global empire.[38]

TOWARD A POLITICAL GEOGRAPHY
OF ENERGY TRANSITIONS

In biophysical terms, an economic process can be seen to rely on a continual flux of energy and matter that it degrades while value is created.[39] Here, energy infrastructure plays an enabling role as it "create[s] the conditions of possibility for rapid and cheap communication and exchange across distance."[40] Historically, the Cuban economy has been integrated in a range of infrastructural arrangements, first as a net supplier of energy and matter to imperial cores and later as a net importer within the international socialist political economy and PetroCaribe, the regional trade alliance spearheaded by Cuba and Venezuela. A situated practice of energy use can only be sustained if geopolitical, market, or other social relations maintain a positive exchange of low entropy across space.[41] Indeed, while energy use requires material inputs and generates material outputs, it also establishes relations between places that influence collective forms of social life and political possibility.[42]

The geographies of energy use can be described in absolute as well as relational terms. In the absolute, thrumming generators, rusting cables, coal-hauling trains, sea-going tankers, home-made solar heaters, smoldering kilns, giant gas liquefaction plants, and emptying canisters are concrete, spatially manifest objects, maintained in landscapes in which they enable energy transfers. When they join together to become the conditions of possibility for something else to happen, they become infrastructure.[43] In the relational sense, infrastructures connect and disconnect, include and exclude. They sustain material and semiotic exchanges and embody a political economic logic that specifies the terms of exchange and access. As Brian Larkin argues, infrastructure provides the "architecture of circulation" through which energy, capital, signs, and social power flow to "generate the ambient environment of everyday life."[44] Analytically, therefore, we can trace the absolute and relational geographies of the socio-ecological circumstances within which an everyday practice of energy use is situated—the circumstances that provide infrastructural form for human life.

The Cuban islands offer an interesting setting for developing a perspective like this. In geophysical terms, an island is an absolute space with its own geological and ecological characteristics and a finite resource base. However, islanders tend to engage in material and symbolic exchanges with populations beyond the territorial limits of the islands they inhabit in order to sustain the reproduction of socioeconomic relations. By engaging in extra-local exchange, they import a surplus of energy and materials that enables them to overcome environmental limits. "Islands may be geophysical givens of material reality," as Eric Clark writes, "but how they develop is determined more by their positions, positioning and relations in an evolving socio-ecological system than by their physical limits or characteristics."[45] The history of revolutionary Cuba is one in which the island population to varying degrees has been able to engage in socio-ecological exchanges with states and capitals beyond its shores. To appreciate the material reproduction of Cuba's island economy over time, then, we need a "progressive sense of place."[46]

Just as an island is a node in an extensive network of socio-ecological exchange, industrial technology can be seen to have a socio-ecological geography. A machine, or a power tool for that matter, works as long as it is sustained by a flux of low entropy harnessed from a local environment or imported from an environment spatially displaced. For Alf Hornborg, machines tend to be perceived as bounded objects in modern societies: technological artifacts of various kinds—a Cuban cane harvester or a Soviet-made electric iron—pay testament to human ingenuity and modernity. But such "machine fetishism" neglects the political geographies that keep machines from "dissipating" as nodes in wider networks of ecological flows.[47] "Technologies are . . . contingent on the ratios at which energy, materials, and other inputs and outputs are exchanged in human societies," Hornborg writes and continues: "In accordance with the Second

Law of Thermodynamics, we know that the output of any technological system will represent less available energy or productive potential than the input required for its production. To be viable, in other words, a technological system must be reproduced through ecologically asymmetric resource flows."[48]

As an intersection more than a bounded object, a "bundle of lines" more than a "blob" to borrow from Tim Ingold, energy technologies run on a continuous supply of low entropy. The flow is less *in* the machine than the machine is *in* the flow.[49] A technical artifact is held in place in a matrix of socio-ecological relations, and as such, it is a social strategy of accumulation. The ability to maintain a large technological capacity rests on the ability to accrue ecological resources and to uphold the infrastructural arrangements that sustain the flow of these. Thus, energy use is contingent on a geopolitical mode of social power that organizes and spatially orchestrates the uneven distribution of resource flows.[50] In this book, I show how the use of energy, in various forms, gives rise to politicized geographical patterns and how key political dynamics of autonomy and dependence are negotiated as these patterns are reconfigured at moments of energy transition.

POLITICAL IMAGINARIES AND THE INFRASTRUCTURAL STATE

To examine the territorial organization of energy flows in Cuba, it is difficult to circumvent the role of the socialist state. Indeed, Cuba's low-carbon contradiction raises fundamental questions about the state in energy transitions and the socializing role of infrastructure. A classic understanding of the link between infrastructure and the state is captured in the notion of infrastructural power: here, infrastructure is a means for the state to wrest social order through the impact of an extra-local socio-technical system. Put differently, a territorialized infrastructural system enables a central government to implement policy across its realm and, as such, it is key to processes of state formation.[51] For example, when decisions over electrification were devolved from the federal to the state level in India after independence from British rule, the postcolonial state expanded its presence, but it was also unevenly territorialized across the subcontinent.[52] In Mozambique, the liberation movement Frelimo extended electricity infrastructure into rural areas after the country's independence from Portugal. This centralized infrastructure enabled new social and economic practices to unfold, and these were practices in which narratives of the postcolonial state were reproduced. As Marcus Power and Joshua Kirshner argue, the state appeared territorially as an effect of infrastructure.[53]

In many accounts of infrastructural power, territory is an uncontroversial notion. It is an irreducible terrain for state power, which reflects a wider tendency in environmental politics to conceptualize the state as an autonomous actor or an institutional realm that regulates, simplifies, and intervenes in the environment.[54] However, building on recent work in political geography that collapses the dualism between state and environment, territory ought to be seen less as an absolute backcloth to state power than an effect of networked socio-technical and socio-ecological practices in which "the state" is performed.[55] When people interact with an infrastructured environment in the name of the state—or do so against or even beyond it—the state is continuously territorialized and reterritorialized. Thus, if technologies and economic activities are nodes in networks of socio-ecological flows, the state emerges when such networks are territorialized in its name at particular scales through everyday practice.[56]

In his recent critique of degrowth, Paul Robbins identifies large-scale infrastructural solutions as central to eco-socialist projects. The product of past energy expenditures and labor, infrastructure creates the conditions of possibility for new, larger-scale forms of community and control to emerge.[57] I argue that the Cuban socialist state in part emanated as an effect of an oil-fired electricity infrastructure—as a vehicle of socio-ecological transformation. The SEN engendered a scale-based imaginary of community, calling on a postcolonial socialist subject to interact with it. Thus, through everyday practice, that subject was brought into being in relation to the SEN. The view of large-scale infrastructure as something progressive sets eco-socialism aside from degrowth proposals that tend to envisage a downscaling of social and economic life, seeking "frugal and egalitarian small-scale societies" in which some readers identify a "tacit anti-urbanism."[58] In reviewing and formulating radical environmental proposals, and working through how they require a reterritorialization of energy flows, we need to pay close attention to the role of the state and the infrastructural practices through which it forms.

Such a perspective on the state adds to scholarship revealing how the technological, cultural, and political orders were coproduced in a range of national contexts through energy development. Notably, a rich set of studies on the history of the Soviet energy industries have emphasized the role of an emerging professional community of system builders, engineers, and scientists whose technical choices shaped not only patterns of industrial organization but also the legitimacy and geopolitical relations of the emerging Soviet Union.[59] However, if "the political" designates how power, in its multiple forms, is exercised and negotiated in human relationships, it is important to look beyond the systems level to also appreciate how energy transitions reconfigure social hierarchies and unequal access arrangements in lived-in landscapes. As Vanesa Castán Broto argues, an energy transition "requires rearranging the mundane environments that confer stability on a specific arrangement of resources, technologies, institutions, and social practices." In other words, a study of infrastructural transformation asks for an expanded sense of what it means to "act politically."[60]

To this end, the notion of infrastructural form is central to this book. The word *form* is significant because it implies both object and process. As a noun, *form* speaks of infrastructure as an absolute physical presence, but as a verb, it captures infrastructure's relational qualities—how it forms social life and "creates a sensory, tactile environment that translates political rationalities into ambient experience."[61] Indeed, for Penny Harvey, infrastructure is an attempt to "create systemic relational form" that enables and forecloses human action.[62] When infrastructure is mobilized as part of a political economic project, it is also co-productive of imaginaries that bring subjects into being. The subjectivities associated with oil-based energy use are particularly significant to understand at a time of global climate change.[63] In the pages to come, I show how an imaginary of energy use was placed in Cuban politics that centered on a postcolonial, class-based subject. However, the special period generated experiences of a kind that were unequally shaped along lines of gender, race, and class. As much as the enforced low-carbon situation led to a proliferation of new infrastructural arrangements, it shook the master narrative of socialist progress, fractured and reterritorialized the socialist state, and required Cubans to negotiate the inequalities they experienced within a wider cultural context of revolutionary change.

AN OUTLINE OF THE BOOK

To unravel the implications of Cuba's low-carbon contradiction for energy transitions, the next chapter examines the broader setting of Cuba's deepening oil dependence in the mid-twentieth century. A nationalist critique demarcated the Cuban islands as spaces fundamentally lacking extractable resources, and in revolutionary discourse, the Cuban islands morphed with a nationalist subjectivity deprived of development potential. The situation of resource scarcity rested on a relational concept of the island's resource base: in the nationalist narrative, the oil majors kept their discoveries in Cuban oilfields secret, the foreign-owned sugar companies had deforested the island, and the uneven development of the gas and electricity industries only reflected the profit motives of their US-based owners. When the Marxist-Leninist faction of the revolutionary movements seized power over the unions in the early 1960s, a socialist interpretation of the need for electrification began to inspire the government's development strategy. The exchange of Soviet oil for Cuban sugar provided the metabolic conditions for the socialist state's formation.

Chapter 2 focuses on the significance of the SEN for the socialist project and how a transition to a particular form of electricity infrastructure embodied a scalar strategy for socialist development. Two basic assumptions underpinned Cuban Marxist debate: first, that increasing energy use was definitive of development, and second, that a centralized, nationally integrated infrastructure was a requisite material condition for social justice—or in more contemporary terms, "energy justice."[64] Together with the state rationing system, the SEN territorialized a political economy of energy production, distribution, and use, which enabled the socialist state to appear as a vehicle of just energy supply. In the 1980s, the Soviet oil industry went into decline, leading the Cuban government to launch a major energy reform. While improved work efficiency was deemed a revolutionary imperative, the centerpiece of the reform program was a Cuban

"Project of the Century." Setting out to build a nuclear power plant, the government reconfigured the infrastructural form of the Revolution, reducing the island's dependence on Soviet oil exports while developing the country's productive forces to the highest level.

The nuclear program came to an abrupt halt at the same time as Cuba lost its beneficial oil import agreement with the Soviet Union. Chapter 3 examines how urban household practices changed in the years following the disintegration of the socialist bloc based on ethnographic encounters in Pinar del Río and Havana. Differentiated by gender in particular, Cubans interacted with energy infrastructure in uneven ways, and their unequal relations to the state's deteriorating infrastructural form stratified experiences of the crisis. The plunge in energy consumption not only generated material inequalities but also challenged key premises of the revolutionary master narrative. Cubans in government and the household setting alike negotiated and normalized their lived experiences by developing contrasting but intertextually working narratives of national siege, resistance, and inventiveness. In these narratives, the metahistorical category of the "special period" allowed them to disarticulate the gendered inequalities that were intensified in the act of energy use, foreclosing a wider ideological crisis.

Moving from the urban context, chapter 4 focuses on responses to the energy crisis in government, academia, and industry. These responses can be interpreted both in terms of an eco-socialist and a degrowth paradigm of attempted energy transition, and I attend to the political priorities associated with the latter in particular. Here, the Programa Energético takes center stage—a government policy adopted in 1993 to guide activity in the state economy. The Programa Energético created a space for the population to set up infrastructural systems parallel to those of the state, enabling them to withdraw from state relations. The new energy systems provided the conditions of possibility for increased local autonomy, but the proliferation of smaller-scale infrastructures and localized spheres of exchange generated political conflict around the commodification of fuels in an increasingly porous state formation. When the infrastructural form for everyday life changed, energy use was a practice in which the political future of socialist Cuba was negotiated.

Chapter 5 examines a critical juncture in the early 2000s, which saw Cuban life transformed by three events: the onslaught of hurricane

Charley, the Energy Revolution, and the emergence of the Bolivarian Republic of Venezuela in regional affairs. While the Energy Revolution drastically decarbonized the economy in relative terms, Cuba's membership in PetroCaribe secured access to oil from Venezuela under nonmarket conditions. Venezuelan oil allowed Cuban energy consumption to increase in absolute terms, sustaining the low-carbon contradiction. Based on encounters in industry and households, I assess far-reaching efforts to reduce the energy intensity of the Cuban economy during the Energy Revolution. These center on an energy transition enacted in Cuban households, intervening in the gendered division of labor, and an ecstatic appeal for energy efficiency in official rhetoric. An ethics associated with energy efficiency and the meaning of being a "true revolutionary citizen" extends from the energy reforms of the 1980s (chapter 2) via the special period (chapters 3–4) to the Energy Revolution (chapter 5). This ethics, expressed in efforts to create a culture of "thrift" (*ahorro*), worked to recentralize power and reinforce the hegemony of the socialist state.

I conclude the book by returning to what Cuba's low-carbon contradiction says about our understanding of energy transitions. In contrast to the Anglo-American experiences that dominate current scholarship in this area, Cuban history invites us to reevaluate key assumptions about energy transitions: what happens if we start from a peripheral Caribbean island instead?[65] Cuban socialism, as it first developed during the Cold War, was firmly inspired by the Soviet example and readings of Marx that emphasized the "development of the productive forces" for economic progress. Interestingly, contemporary eco-socialist debates, in Cuba and internationally, approach low-carbon development in a manner that echoes core aspects of these earlier discussions. Nevertheless, the postcolonial vantage point exposes a crucial geopolitical dynamic: historically produced power relations shape the conditions of energy use so that the possibility of bringing about eco-socialism or other growth-based visions is contingent on a population's position in the global political economy. While large-scale infrastructural solutions continue to dominate in Cuba, degrowth-inspired practices are also integral to island life. Conflicting political priorities of autonomy and centralization, self-sufficiency and redistribution are held in suspension in the decarbonization of the economy.

1 Against the Energy Empire

When Fidel Castro introduced the first Agrarian Reform Law in May 1959, described as the Revolution's "basic law," he appointed Antonio Núñez Jiménez to oversee its implementation as head of the National Institute of Agrarian Reform (INRA). Núñez Jiménez was a captain in the rebel army and known widely for his work as a professional geographer. While managing the land reform, Núñez Jiménez also found the time to adapt his textbook *Geografía de Cuba* to the country's new "revolutionary" high school curriculum. In the foreword, Minister of Education Armando Hart Dávalos explained that the first edition had been confiscated and burned on the orders of President Batista. Cuba's physical geography was a controversial matter. Both Cuban and foreign geologists were certain that "the precious black liquid abounds in Cuba," Núñez Jiménez wrote. Every government since the dictatorship of Gerardo Machado had claimed to be working for the industrialization of Cuba on the basis of oil, but despite decades of exploration, the oil discoveries had been few and small. When the foreign oil companies had prospected for oil, they had either kept their discoveries secret or sat on the pumps in the fields they had found. "The interest of these foreign companies," Núñez Jiménez argued, "is to make it appear as if no oil exists in Cuba with the idea of keeping our expected deposits in reserve in case of war." As a result, the Cuban people had been forced to import oil from the oil majors in its pursuit of national development.[1]

The belief in this conspiracy went decades back. The historian Eric Gettig documents how two competing narratives emerged around oil in the first half of the twentieth century. Foreign oilmen, backed by the United States, insisted that discoveries were around the corner and portrayed themselves as enthusiastic partners in the modernization of the Cuban economy. Frustrated by the small finds, by contrast, Cuban nationalists such as Núñez Jiménez argued that the foreign companies hid Cuba's oil wealth for their own future benefit.[2] The oil majors had indeed gone to great lengths to "produce scarcity" in the Middle East, seeking to establish a stable market and maintain profitability in an industry prone to overproduction. By forming cartels and controlling strategic chokepoints in the commodity chain, they ensured a shortage by restricting output, thus raising prices.[3] More directly, however, the two Cuban narratives should be interpreted in the regional context of the Caribbean. Until the 1960s, global oil production centered on the coasts of the United States, Mexico, and Venezuela, where the world's greatest oil reserves were found. These deposits encircled the Caribbean island-states that barely had any known reserves at all. Given the booming industries to its north and south, it was hardly unreasonable to think that Cuba also must be endowed with great oil wealth.[4]

In contrast to the notional geography of oil, several features of the Cuban landscape had developed over the *longue durée* with some certainty. The Cuban archipelago is today made up of the Cuban main island, the smaller Isla de la Juventud (Isle of Youth), and a series of islets, known as keys (*cayos*), along the main island's shores. The islands are generally lowland apart from three hill areas on the main island: the Guaniguanico hills in the west, the Escambray Mountains in the center, and the Sierra Maestra in the east. Located in the tropics, Cuba gets ample sunlight, but the islands are also warmed by the ocean current that flows into the Caribbean Sea from the South Atlantic. The current turns around Cuba's western tip into the Gulf of Mexico where it forms the Gulf Stream, thrusting water into the North Atlantic through the Florida Straits. When the ocean is at its warmest, usually from July to October, the trade winds give rise to tropical storms and hurricanes. These weather systems form over the Atlantic and turn northward as they reach the Caribbean, frequently leaving the Cuban islands in their path. The rivers in Cuba swell during this rainy season and turn into small streams when it is dry.[5]

More than a backdrop to human life, however, the Cuban landscape is a product of it. Two principal features can be traced to the colonial history of the past five centuries. First, the landscape is marked by the long presence of sugar cane. Sugar was unknown to the Caribbean until the early sixteenth century when it was introduced by the Spanish. After the Haitian Revolution of 1791, Cuba superseded Haiti (then Saint Domingue) as the world's leading sugar-exporting colony. Sugar monoculture became the all-dominating feature of the Cuban economy and sugar cane the all-dominating species in the island's ecosystems.[6] Second, Cuba has little forest cover. As Reinaldo Funes Monzote demonstrates, pre-Columbian Cuba was densely forested, but Cuban wood was both ideal shipbuilding material for the Spanish Armada and abundant fuel for the burgeoning sugar industry. Newly cleared land also provided fertile soils. When the soil eroded after a series of cane crop cycles, new land was cleared to the same effect. The naval and agricultural interests competed for wood until 1815 when the Spanish Crown granted sugar planters the right to clear land at will. This led to rapid deforestation, providing land for cane and energy for the mills.[7] "The death of the forest," the historian Manuel Moreno Fraginals later lamented, "was also, in part, the long-term death of the island's fabled fertility."[8]

As the largest, centermost island in the Caribbean, Cuba was an important hub in the regional trade networks prior to the Spanish conquest. For the European powers, it was also important to control Cuba, and particularly Havana on the north coast, since it secured the fastest sailing route from the Americas to Europe along the westerlies.[9] The port of Havana long held an exclusive royal privilege for commerce, which made it Cuba's main economic center.[10] The province of Pinar del Río developed somewhat in the shadow of the capital. Unlike most other parts of the country, the sugar industry only partly expanded into this westernmost region, which instead became the heartland of Cuban tobacco farming. In contrast to sugar production, well suited as it is to mechanization and centralization, tobacco was and still is grown by small farmers almost exclusively.[11]

The place-name Pinar del Río ("pinewood by the river") speaks to the existence of forest cover, at least at one time, and especially the endemic Caribbean pine (*Pinus caribaea caribaea*) and tropical pine (*Pinus tropicalis*).

By the mid-nineteenth century, Spanish influence in Latin America was waning and the economic and cultural relations between Cuba and the United States tightening.[12] US capital fast entered the sugar industry, and by the 1870s, 75 percent of all Cuban sugar was destined for the United States.[13] Havana quickly became a wealthy modern city for the Spanish and US elites, served by incomes from its sugar-producing hinterlands.[14] This was reflected in the city's as well as the country's uneven infrastructural development. Early investments in gas and electricity attested to the close-knit relations forming between Cuba and the United States. Lanterns and kerosene lamps were long used to provide streetlight in Havana, but in 1844, a group of US businessmen received a royal concession to build a coal-gas plant with underground tubes and streetlights. In the name of the Compañía Española de Alumbrado de Gas (Spanish Gas Lighting Company), the business set up a plant in Tallapiedra from where they operated Cuba's first public lighting system with centralized energy supply.[15]

While the gas industry expanded on the island, experiments with arc and incandescent lighting were taking place in the United States. Thomas Edison and other inventor-entrepreneurs also carried out experiments in Havana, and in May 1882, Edison installed an electricity system with incandescent lighting in the café "El Louvre" overlooking Havana's Parque Central (Central Park), demonstrating the commercial importance of Havana to the US elite. In September that same year, he inaugurated his famed first electricity system in New York's Pearl Street.[16] In 1888, the Tallapiedra plant was repurposed to lay the foundation for Havana's first permanent electricity system with arc lamps replacing gas lanterns on the city's main commercial streets, Obispo and O'Reilly. From this point onward, electrification proceeded relatively quickly. In cities across the island, privately owned utilities invested in urban supply systems, initially providing streetlight and public transport via tramlines. In the city of Pinar del Río, a local enterprise called La Industrial (The Industrialist) installed a public electric lighting system in 1893.[17]

At this time, a protracted anti-colonial struggle unfolded in Cuba. In the revolutionary rendition of the struggle, the Ten Years' War (1868–78) and the War of Independence (1895–98) were its most critical moments. Cuban nationalists such as José Martí, Antonio Maceo, and Máximo Gómez led their forces into battle with the Spanish Crown. Later, Fidel Castro asserted that the guerrilla war waged by his 26th of July Movement (M-26-7) was a direct continuation of this anti-colonial struggle. The Cuban Revolution thereby testified to the Cuban people's relentless struggle for liberty and justice, with revolutionary discourse fashioning a collective national identity of resistance.[18] In 1898, the United States intervened in the Cuban war for independence, ultimately leading Spain to relinquish Cuba alongside Puerto Rico, the Philippines, and Guam to the new American world power. Four years later, the United States declared Cuba independent, but under the Platt Amendment, the new constitution granted the US government the right to intervene in Cuban internal affairs.[19]

US investments in the Cuban economy soared and not least in the electrical industry. More widely, utilities in Europe and the United States shifted capital from domestic markets in the early twentieth century, seeking to expand operations by creating markets abroad.[20] General Electric invested heavily in Latin America, where it acted through its subsidiary Electric Bond and Share. The subsidiary in turn placed its acquisitions in the firm American and Foreign Power Company (AFP). After purchases in Panama, which put the energy infrastructures servicing the Panama Canal under US control, AFP turned its attention to Cuba, where it gained the personal support of President Machado. According to one report, General Electric put up eight million dollars for initial purchases in Cuba in 1922. Two years later, AFP had invested another four million dollars in equipment and construction, and by 1932, it had committed over one hundred million dollars on the island.[21] As a result, Cuba's manifold electricity systems merged into two large transmission networks—the Oriental and the Occidental Electricity Systems. When AFP then joined its Cuban acquisitions in the newly formed Compañía Cubana de Electricidad (Cuban Electrical Company) in 1928, it established a complete monopoly over the two networks and, hence, urban electricity supply. In 1958, 56 percent of the Cuban population had access to electricity through the Compañía Cubana de Electricidad's two networks.[22]

The sugar companies also invested in electrification. In the late 1950s, there were more than sixty independent electricity systems based in and around Cuba's sugar mills. The energy was used in the *centrales*, as the mills are known, and in the adjacent *bateyes* (company towns) to provide electric light and to power the cranes, elevators, and other mill machinery that increased the rhythm of production.[23] It also enabled transports of sugar to the ports via electrified railroads. One notable route linked the Hershey mill in Santa Cruz del Norte with the harbors in Havana and Matanzas. The railroad, which still operates today, was electrified by the US chocolate-tycoon Milton Hershey in 1920, and while allowing local passenger and goods transport, its main purpose was to sustain the supply of sugar to Hershey's chocolate factory in Pennsylvania (ironically, a factory today branded as "The Sweetest Place on Earth").[24]

The historian Louis Pérez argues that commercial success in the Cuban electricity sector largely depended on the creation of an urban market for domestic appliances. For utility-scale investments to be profitable, there had to be a growing demand for electrical energy, and since the sugar industry only worked on a seasonal basis, the utilities tried to recreate the consumer culture developing around electricity in the United States.[25] To Cuba's north, the utilities were following a "grow-and-build" strategy in which capital expenditure rather than sales revenues determined profitability. This meant that the utilities had to invest in new infrastructure in anticipation of demand increases to remain profitable. As a result, it was also necessary for the utilities to "produce electricity consumers," as Conor Harrison argues, to keep up with the rate of investment, which they did by marketing domestic appliances within narratives of "progress," cleanliness, and increasing safety.[26] When new appliances became available in the United States they also often appeared in the Cuban market.[27] As a sign of the times, it is fun to note that Wormald, the protagonist of Graham Greene's *Our Man in Havana* (1958), works under cover as a vacuum cleaner salesman, selling brands with head offices in London and New York. A 1954 report from the International Bank for Reconstruction and Development establishes the importance of growing consumption for continued electrification most succinctly: "Cuba's growing population, and also increasing uses of electricity for refrigeration and air-conditioning will undoubtedly mean increasing demand for power over the next few years."[28]

Well into the first years of the Revolution, news reports and political proclamations in the M-26-7's newspaper *Revolución* were interspersed with advertisements for refrigerators and toasters from US companies.

The advertisements all appealed to a strongly gendered housewife ideal. The kitchen was a domestic space in which the demand for electricity could increase quickly through the introduction of new appliances. When *Revolución* rehearsed the Revolution's achievements on its first anniversary, an advertisement depicting a woman pointing to the GE logo reminded the readers that "General Electric is the [company] that best of all knows the things of benefit to the housewife."[29] In time for Mother's Day 1959, the Compañía Cubana de Electricidad ran an ad explaining that nothing would make a mother cherish her children more than the gift of an electrical appliance: "The greatest thing for mother is the love of her children; and it is the children, with their signs of affection, who make the home where she reigns a happier place. / She deserves the best! / How happy you will make 'Mom' if you give her an electrical appliance! Remember that with electricity you live better, and the wellbeing of the family in the home is the greatest happiness for a mother. / Cía Cubana de Electricidad."[30] Next to a notice about an event held to celebrate "the martyrs fallen under the Batista dictatorship," another ad noted that General Electric's products were "an Index of Progress."[31] On yet another occasion, the Compañía Cubana de Electricidad capitalized on the sense of postcolonial triumph in Cuba. Aligning itself with "the movement of patriotic nationalism in which our Fatherland lives," it marketed "Stoves and Water heaters made in Cuba for *Cuban* workers." The appliances were assembled in Cuba, but the components were manufactured by Westinghouse.[32] The cultural and economic relationship between Cuba and the United States was so firm it was even technically built into the Cuban energy infrastructure. In the 1950s, Cuba's distribution systems operated at 110 volts, 60 Hertz alternating current, following US industrial standards. This standard was only suitable for US appliances, and as Pérez concludes, "Much in the Cuban sense of future and of place in that future was shaped by or otherwise derived from the encounter with the North."[33]

STRUCTURAL ECONOMICS AND A SOCIALIST TURN

The electricity systems were concentrated in urban areas and especially in the entertainment districts and the residential neighborhoods of the white upper class. In the nationalist critique of the "pseudo-republic"—Cuba under the rule of Fulgencio Batista—this uneven infrastructural geography played into a broader anti-colonial analysis put forward in the 1950s.[34] Years past political independence, most Latin American countries still depended on primary commodity exports for their national income while at the same time relying on foreign capital for investments and industrial imports for consumption. The United Nations' Economic Commission for Latin America (ECLA) argued that the global capitalist economy was developing in a geographically contradictory fashion, giving rise to two interdependent regions. The Latin American economies belonged to the global periphery, which served the Euro-American core with raw materials, labor power, and export markets. This unequal structural relationship generated increasing development in the core but perpetuated the underdevelopment of the periphery. As primary goods had a lower income elasticity than industrial goods, "The poorer countries could never catch up or compete because they lacked technology and capital."[35]

The analysis became known as the Prebisch-Singer thesis after ECLA's Executive Secretary Raúl Prebisch and the economist Hans Singer.[36] For ECLA, the antidote was a development strategy known as import substitution industrialization, implying that nations in the periphery should raise trade barriers on industrial goods to gradually replace them with domestic industry.[37] In his revised edition of *Geografía de Cuba*, Núñez Jiménez rehearsed the argument by developing a relational understanding of the Cuban islands. To "sustain the simplistic idea that 'Cuba is an Island,'" he wrote, was to misrepresent the country's geography. Cuba had been incorporated in the periphery of the world economy to export sugar to the core, first under Spanish colonialism and later US neocolonial rule.[38] In the early twentieth century, foreign capital controlled the industrial infrastructures in Cuba—from the electricity systems to the telephone lines and the railroads. "The electricity belonged to the Compañía Cubana de Electricidad," as a senior Cuban historian explained when I interviewed him in 2015, "but it shared nothing with Cuba more than the name." Nevertheless, as an epoch-making event, the Revolution would release the national economy from the bonds of the neocolonial market, allowing Cuba to develop.

When the M-26-7 entered Havana on January 1, 1959, it was still unclear what this development entailed in practical terms. Two broader perspectives divided the revolutionary movements. On the one hand, there was a softer faction of nationalist revolutionaries represented by David Salvador, leading the M-26-7's clandestine trade unions, and the popular *comandante* Camilo Cienfuegos. They sought a return to the 1940 constitution with multi-party elections, import substitution, and Cold War nonalignment. On the other hand, a Marxist-Leninist faction argued that ECLA's approach to industrialization had to give way to socialist economic models. The most vocal Marxist-Leninist leaders—Ernesto Che Guevara, Raúl Castro, and members of the Moscow-faithful Partido Socialista Popular (PSP—Popular Socialist Party)—argued for the introduction of a vanguard party, economic planning, and agrarian reform with collectivization. Sneeringly, their opponents referred to them as *los melones* (the melons): they were green on the outside (the color of the M-26-7) but red on the inside (the color of communism).[39]

While the M-26-7 had its support base in rural areas, underground trade unions organized anti-Batista groups in cities. The official Confederation of Cuban Workers (CTC) organized a large part of the workforce but supported Batista under the leadership of Eusebio Mujal. Within CTC, the Electrical Workers' Federation was a particularly influential union which had made considerable gains through collective bargaining. The electricians had played an active role in the Revolution of 1933, which saw the end of Machado's rule, when after a strike and a consumer boycott, they seized control over the Compañía Cubana de Electricidad for about a month.[40] Once Batista and Mujal fled the country in 1958, the revolutionary movements quickly seized power over the CTC. When the union met for its tenth congress in November 1959, the nationalist revolutionaries gained strong support, especially among the electricians.[41] This led Fidel and Raúl Castro and Minister of Labor Augusto Martínez Sánchez to intervene, late at night, to secure both M-26-7 and PSP influence. In the months following the congress, the *melones* in CTC's new executive committee started ousting more moderate nationalists within the elected union leadership. For the time being, however, the leaders of the Electrical Workers' Federation withstood the purges with strong support among the membership.[42]

Meantime, the revolutionary government decided to drastically lower the electricity tariff. The decision coincided with a government investigation concluding that the Compañía Cubana de Electricidad was heavily "overvalued," and that the utility had to reduce its expenditures while continuing to electrify the country.[43] The lowered rates stoked the conflict among the revolutionary movements. Critical electricians claimed that the government was attempting to lower their salaries as part of the Compañía's restructuring. Acting in the interest of the working class, the revolutionary leadership should respect the terms of the electricians' collective-bargaining agreement.[44] The leaders of the nationalized utility responded that the electricians' worries were unfounded and that they instead should focus on the task of electrifying their now independent country.[45] An article in *Revolución* pointed out that the electricians had to consider the interest of the working class as a whole, rather than assert their own short-term interest.[46] Two distinct rationales dividing the labor movement underlay the disagreement: unionization and syndicalist tactics internal to the mode of production versus a socialization of the means of production in a qualitatively new workers' state.

The conflict culminated on November 30, 1960. In the wee hours, a group of electricians blew up strategic intersections connecting the electricity systems in Centro Habana and Tallapiedra, causing a large part of the capital to blackout. When the saboteurs afterward tried to escape to Florida, they were captured by the Revolutionary Maritime Police. In a meeting later staged in CTC's Workers' Palace, Fidel Castro read from a report drafted by the G-2, the M-26-7's secret police, linking the saboteurs to the leaders of the Electrical Workers' Federation (Figure 2).[47] While *Revolución* published statements from other workers' federations calling on the electricians to "clean your ranks," "put the terrorists and saboteurs up against the wall," and make "the fatherland sell-outs [*los vende-patria*] face the avenging rifle of the workers' militias," the government seized the opportunity to oust the union leaders, notably Amaury Fraginals and Fidel Iglesias.[48] This secured Marxist-Leninist control over the CTC.

ENERGY USE AND THE LENINIST EXAMPLE

While restructuring the electricity sector, the revolutionary government began to transform the Cuban economy broadly and rapidly. At first, agricultural production would be diversified under the so-called anti-sugar policy, and with ECLA advisers present in Havana, industrialization would

Figure 2. Compañía Cubana de Electricidad. On the day of Fidel Castro's appearance in the CTC, *Revolución* reported that workers in the nationalized Compañía Cubana de Electricidad congregated in front of the company headquarters to condemn the saboteurs: "The public, in the streets and from balconies, supported them [the workers] with ovations and the shouting of revolutionary slogans." Architecturally and functionally state-of-the-art, the Compañía's building was completed in 1958, and as the architectural historian Eduardo Luis Rodríguez argues, the aluminum-clad façade and open floor plan represented "an undisputed milestone in Havana's modernity." Today, it is the seat of the Ministry of Energy and Mines—formerly the Ministry of Heavy Industry. *Sources: Revolución,* "Ratificarán esta noche los obreros eléctricos su apoyo a la Revolución," 8; Rodríguez, *The Havana Guide,* 198. Photo by Gustav Cederlöf.

take off through import substitution. However, it proved difficult to access the raw materials the country needed in its imported factories, and sugar exports could quickly generate revenue in the short term.[49] To coordinate its efforts, the leadership convened a First National Meeting on Production in Havana in August 1961. The meeting decided that the country would return to a focus on sugar production in the short term to enable industrialization in a number of priority areas in the long term.[50] Addressing the plenary in his capacity as minister of industries, Che Guevara emphasized the investments the government had made in areas ranging from cement and metallurgy to sugar, synthetic textiles, and pharmaceuticals.

But he also noted that the expansion of these industries would increase the demand for electricity. To enable industrialization, large-scale electrification was essential.[51]

Economic models from the socialist bloc soon replaced ECLA's development economics in Havana. When the magazine *INRA*, published by the eponymous government agency, ran a piece on the Soviet electrical industry across three full spreads in June 1961, it suggested that Cuba had to learn from the Soviet experience in order to develop as a nation. It was necessary to follow the Soviet example to fulfil "the primordial revolutionary task of creating the techno-material base of socialist society."[52] In Soviet state orthodoxy, electrification had attained an almost mythical status: it was an imperative in the historical transition to communism. In the Leninist reading of Marx, communism denoted a society of material abundance in which everyone had access to the material goods they needed ("from each according to his ability, to each according to his needs"). Material abundance, in turn, was a question of increasing economic productivity. More complex technology, an ever more skilled workforce, and more efficient management techniques—in Marx's words, more highly developed "productive forces"—would increase the levels of production and reduce the labor effort needed in the economic process. History, in this reading, was a cumulative process of increasing productivity in which technology and knowledge gained at an earlier stage benefited the dominant class interests of a later, more advanced stage.

In a famous passage in *The Poverty of Philosophy*, Marx argues, "In acquiring new productive forces men change their mode of production; and in changing their mode of production . . . they change all their social relations." He then arrives at a conclusion making it seem as if social relations spring out of machines: "The hand-mill gives you society with the feudal lord; the steam-mill, society with the industrial capitalist."[53] In his work in the years around the Bolshevik Revolution, Lenin found that steam had been the driving force behind the development of capitalism in Britain, and he argued vehemently that electricity would provide the "techno-material base" of socialism.[54] New blast furnaces, steam boilers, and conveyor belts in an emerging industrial economy required a continuously increasing supply of electricity. To *INRA*, this was abundantly clear from the Soviet example:

The Soviets need ever more electrical energy, to set up new industrial companies and to improve the old ones with automated workshops and factories, for the electrification of the railroads and for the [electric] lighting of the cities and villages, of the *sovkhozes* and *kolkhozes*.

Metal, fuel, and machinery are the *base* of the socialist industries. Without metal, it is impossible to manufacture anything in any branch of modern industrial production. Millions of tons of steel, aluminum and various alloys are today melted in *electric furnaces*. It is more convenient and more profitable. However, to obtain one ton of aluminum, for example, it is necessary to consume 20,000 kilowatts, per hour, of electrical energy.[55]

To develop the new Soviet society, Lenin appointed a State Commission for the Electrification of Russia (GOELRO) in 1920. Only Moscow and Petrograd (St. Petersburg) had limited access to electricity through two urban utilities at a time when most of Russia was in ruins after the Civil War. In stark contrast to this environment, GOELRO proposed an integrated all-Russian electricity network extending into the countryside. Twenty-seven regional power plants would energize the centralized network, acting as concentrated nodes of generation.[56] The plan seemed so distant even its chief architect, Gleb Krzhizhanovsky, questioned its viability.[57] But Lenin pressed on. Electrical cables would bring industrial productive forces to the peasantry, proletarianizing them, and the new techno-material base would undermine any capitalist tendencies in the countryside. In raising poles and drawing cables, the Revolution's electricians waged a class war against the peasants. National electrification would secure the socialist relations of production.[58]

When Lenin presented the GOELRO plan to the Eighth All-Russian Congress of Soviets, he announced that "Communism is equal to Soviet power plus electrification of the whole country."[59] As Anindita Banerjee points out, electricity took on a dual meaning with Lenin's maxim, which later would echo through Cuban energy policy. Electricity became both a material base and a metaphor in Leninist thought through which the electric current literally and figuratively brought the masses out of the darkness and into the light. Electricity generated heat, light, and movement, but it was also "an extension of abstract political authority (Soviet power) and a catalyst for epistemological change (Communism)."[60] Electrification was conducive of infrastructural form: it extended the spatial reach of the workers' state, gave rise to more developed productive forces, and made way for the automation of production. Lenin's declaration of the New Economic Policy (NEP) in 1921 forestalled the implementation of the GOELRO plan, but it was incorporated in Stalin's first five-year plan in 1928. Lenin's grand vision to electrify Russia then turned into an origin myth in Soviet orthodoxy, becoming celebrated as the source of the Soviet Union's historical achievements and showcasing the possibilities of centralized planning.[61]

In 1963, Fidel Castro returned to Havana after his first visit to the Soviet Union, and he immediately announced that it was right for Cuba to focus on large thermoelectric plants. The size and the economies of scale they generated made them the most efficient form of production.[62] Hardly by accident, large power plants were the kind of infrastructure he had witnessed in Russia, for example when he visited Siberian Bratsk to behold the construction works of a 4.5 GW hydroelectric dam on the Angara River.[63] The significance of centralized infrastructure could not be exaggerated. In *Capital*, Marx shows how capital must be immobilized in the landscape in the form of machines, factories, and other fixed assets to produce surplus value. While labor is the source of value, fixed capital is the medium and the impetus for its expansion. When the productive forces developed, the means of production would become more strongly concentrated in space to facilitate the expansion of capital.[64] Che Guevara argued that the 1 GW generators existing in the United States exemplified this process, having an immense capacity concentrated in one machine. A single 1 GW generator could produce more electricity than all of Cuba's power plants together with a fraction of the human labor input. That these generators existed in the United States but not in Cuba demonstrated how the US economy had reached a higher stage in the development of the productive forces.[65]

Just like the Soviet government, Cuba's revolutionary leadership identified the relationship between electrification and industrialization as one of causality. They spoke of electrification as if the electric current had an inherent social force, inevitably bringing industry in its wake. The potential for industrial production lay dormant in the country awaiting the electric current to bring it to life. In a speech to celebrate Antonio Guiteras, a leading socialist politician during the Revolution of 1933, Che Guevara argued that to control the supply of electricity was to control "the rhythm" of industrialization.[66] Reflecting on the importance of automation in a meeting in the Ministry of Industries (MININD), he concluded, "We have to think of electronics as a function of socialism and the transition to communism."[67] When the nationalized Compañía Cubana de Electricidad hosted a First Forum on Electrical Energy in Havana in 1963, Guevara invoked the GOELRO plan directly as a model for Cuban development. With the benefit of hindsight, he argued, it was evident that Lenin had overemphasized the importance of electricity itself, but history had shown that electrification provided an indispensable techno-material base along-side mechanization, chemistry, and a deepening of social conscience in order "to pass to higher stages in the development of society."[68] Instead of leaving the development of this indispensable techno-material base to the logic of capital—to the haphazard law of value and the consumer market—the state had to guide electrification consciously from the start.

From a Cuban Marxist perspective, historical progress was tantamount to a twofold structural transformation of the national economy. On the one hand, production would transition from an agrarian to a manufacturing focus, from the machete to the cane harvester. On the other, an organic economy, sustaining its low entropy needs with muscle power and harvests from eco-productive land, would metamorphose into an electric economy. The organic economy drew its energy needs horizontally, but the electric economy was vertical, taking energy from subterranean stocks. While E. A. Wrigley distinguishes between an organic and a mineral energy regime to conceptualize the transition from water to steam power in Britain, electricity was the desirable energy source in Cuba rather than mineral fuels per se.[69] Electricity infrastructure provided the enabling environment for socialist development.

The transformation from an organic to an (oil-fueled) electric economy was linked to a vision of human freedom. When the productive forces developed, machines would take over human work tasks at a continually increasing rate. Under capitalism, this meant that the machines dominated the working class with increasing intensity—automation forced workers to sell their labor for ever-lower wages when they competed with the machines for work. But under socialism, the workers would own the machines collectively and put them to use in their own interest. Mechanization and automation would then be the sources of emancipation, eventually making human labor superfluous to production.[70] In a treatise on energy policy and socialist development from the late 1980s, the Cuban economist Hugo Pons Duarte also defines the relation between worker and machine as one of the fundamental contradictions of capitalism. Yet this contradiction, he argues, holds within it the potential for socialist progress.[71] A vision of liberation thus meshed with a form of energy use, and energy use was inseparably linked to the historically defined social relations that sustained it.

THE GREAT DEBATE AND INSTITUTIONALIZATION

The deliberations on a Cuban energy transition took place within a more general discussion on the Revolution's future direction. This was no longer a discussion between nationalists and Marxist-Leninists but one between different factions of Marxists, pitting Moscow-faithful PSP members against Che Guevara in particular. Later described as the Great Debate in Cuba, the dispute centered on the interpretation of Marx's philosophy of history and its implications in the postcolonial nation.[72] In *Capital*, Marx goes on to argue that market competition forces capitalists to develop more advanced technology and knowledge in order to minimize the costs of production (to extract "relative surplus value"). By spurring innovation and scientific discoveries, capitalism develops the historically necessary conditions for a society of material overabundance. However, if profit, as Marx argues, originates in the capitalist's ability to pay his workers less than the value they add to the production of commodities, a higher degree of automation will lead to production at a falling rate of profit. This gives rise to another contradiction: while the individual capitalist must automate production for the business to remain profitable, increasing automation undermines the ability of the capitalist class as a whole to profit from production. At the heart of capitalism, then, is a contradiction by which the system throws itself into periodic crises. At this time, for Marx, a highly coordinated proletariat—worker coordination being a key productive force—can seize power over the means of production in a socialist revolution.[73]

The question at the center of the Great Debate concerned the need for this process to unfold naturally. Was a period of capitalism necessary for an economy to mature to the extent that the transition to a hyper-productive communist society was possible? In Soviet Russia, a communist vanguard had seized power on behalf of the working class before capitalism had been fully established and organically developed the country's productive forces. In Cuba, the M-26-7 had swept to power while the country was in a state of underdevelopment. The Cuban economy had developed to serve the world industrial core, distorting the development of its productive forces. So, did Russia and Cuba have to develop a modern capitalist economy during a transitory phase, even if it ran counter to the short-term interest of the working class?

From a strictly technical point of view, the Cuban answer was no. Che Guevara argued that Cuba would be able to speed through the technological development already achieved in the industrialized countries by importing machinery from overseas. Though essential for socialist progress, technology itself was politically neutral and separable from its social context. The key difference between capitalist and socialist development was the social relations into which that technology and its users entered—not the materiality of the machines.[74] In Puerto Rico, for example, which was ceded to the United States at the same time as Cuba, electrification also proceeded based on oil-fired thermoelectric plants. Materially, this was the same technology as that used in Cuba even as Puerto Rico was a US dependency.[75] The means of electrification did not change after the socialist revolution, but the visions of where that materiality led Cuba diverged from those predating the Revolution and the visions dominating in Puerto Rico under neocolonial US rule.

Nevertheless, the issue of technical standards was a question of practical importance to Cuban planners and engineers. The Cuban electricity systems were adapted to US standards with appliances running on 110 V and plugs with flat pins. Soviet appliances instead ran on 220 V with plugs using circular pins. The Cuban solution was highly pragmatic and reflected the conviction that social relations rather than technical standards mattered for development. The electricity systems continued to be developed following US standards, bearing in mind that most Cubans had US-made appliances in their homes. Practically speaking, it was also possible to run appliances adapted to 220 V on a 110 V current, as opposed to the other way around. Over time, however, the nationalized utility developed two parallel distribution systems in many parts of the island, which made it possible to run Soviet-manufactured air conditioners and refrigerators on the higher, more efficient voltage standard. In Cuban households today, one socket adapted to the US standard and one to the Soviet are often fitted adjacent to each other.

Looking beyond technology to political economic management, Soviet policy provided one answer to the question at the core of the Great Debate. In the early 1960s, the Soviet economy incorporated several capitalist mechanisms to develop the productive forces. Since its inception, the NEP had been seen as a step backwards for the socialist revolution, as it opened up spaces for free enterprise. When Lenin died, Stalin also forcefully centralized production, yet he still claimed that the law of value—that is, the allocation of labor and resources by market forces—operated in the Soviet Union. Stalinist central planning made use of credits, material incentives, and nonstate property such as cooperative-owned farms (*kolkhozes*) to accomplish plan targets.[76] While Stalin cracked down on debate that challenged his economic methods, the de-Stalinization process led to renewed debate on these matters. Nikita Khrushchev soon sanctioned experiments allowing companies to assess performance based on profitability, and these experiments ultimately resulted in the Kosygin Reform of 1965, which introduced profits, prices, and interest rates as levers in the Soviet economy. With Leonid Brezhnev and Alexei Kosygin at the helm, having ousted Khrushchev, the Soviet leadership hoped that these mechanisms would increase productivity and improve product quality to better meet consumer demand.[77]

In Cuba, the revolutionary leaders followed the discussions in the Soviet Union closely. Carlos Rafael Rodríguez, the head of INRA since 1962, was a member of PSP and argued that the Cuban Revolution should follow the Soviet example. Che Guevara took up a contrary position. He argued that the Soviet Union's economic system originated in the historically contingent NEP rather than objective Marxist science. Furthermore, not only had the productive forces yet to develop for a society to reach communism, but also the social values that characterized communist life. Guevara contended that while the law of value and the competition it spurred developed the productive forces, these capitalist mechanisms would re-produce a culture of individualism and self-interest. The transition to communism required the development of a generous, self-sacrificing mentality. Thus, the socialist state had to undermine the law of value and encourage voluntarism so that the population eventually would work out of social duty rather than material interest.[78]

Guevara also developed an alternative economic management system to the one used in the Soviet Union. In his model, all property was universalized in the socialist state. Just as Lenin before him, he envisioned the national economy as one integrated factory. For the law of value to operate, there had to be a transferal of ownership between independent legal entities, but when different branches of the state exchanged goods among themselves, ownership did not change. A nationally integrated state-economy therefore undermined the law of value.[79] As a result, the universalization of property in the state also de-commodified all goods. This allowed the state to distribute goods as use values to the population in return for labor-time in the state economy. By extension, money stopped being a measure of exchange value and became a means of accounting, upholding reciprocal relations within the national enterprise.[80] Together, these processes allowed the workers' state to allocate labor and resources and to control the surplus product "consciously" instead of relying on the law of value. As an alternative to competition, therefore, the state had to stimulate the development of the productive forces by educating the workforce and rationally planning investments in electrification and automation.

The two competing models were put to work at the same time in the Cuban economy in the mid-1960s: the Soviet-inspired model in INRA and the Ministry of Foreign Trade and Guevara's model in MININD. The debate then came to an end when Che Guevara left Cuba in 1965. Although Guevara's perspective at first lost ground, Fidel Castro soon announced a sweeping reform that drew on his ideas. In 1968, the government nationalized all nonagricultural activities during the "Revolutionary Offensive," appropriating more than fifty-five thousand private businesses and turning them into state property. Throughout 1969 and 1970, Fidel exhorted the population to do voluntary work in the countryside, cutting cane and harvesting coffee. Voluntarism would build the socialist economy and the ethos of the New Man simultaneously.[81] The efforts peaked in 1970 when

PODER POPULAR   PARTIDO COMUNISTA DE CUBA

President / First Secretary
Executive Committee / Politburo
Council of Ministers / Nation \ Secretariat
Council of State / Central Committee
National Assembly /

First Provincial Secretary
Provincial Assembly / Province \ Provincial Executive Buro
Provincial Committee

First Municipal Secretary
Municipal Assembly / Municipality \ Municipal Executive Buro
Popular Council / Municipal Committee

Constituency / Residence/Workplace \ Party Cell

Figure 3. Cuba's political system, 1976–2019. Since 1976, the Cuban socialist state is governed through two organizations, both reflecting the Leninist ideal of democratic centralism: the Poder Popular, in which the National Assembly holds formal legislative powers and the Council of Ministers holds executive powers, and the Communist Party, whose role is to "guide" the state politically. A constitutional referendum in 2019 modified the structure slightly.

Fidel Castro announced that the country would harvest ten million tons of sugar in one year, largely based on voluntary labor. This was more than a doubling of the *zafra* (harvest) from 1963. Despite the obsession with sugar, the Ten Million Ton Harvest failed, and the all-embracing focus on one crop led to a neglect of other economic sectors with the urban workforce mobilized in the countryside.[82] During the 1970s, the guevarista ideas behind the Revolutionary Offensive therefore gave way to an economic model inspired by the Kosygin Reform, known as the System of Economic Management and Planning (SDPE). The revolutionary project also went through a process of "institutionalization" with Soviet-inspired political institutions codified in a new constitution (Figure 3).[83]

ENERGY, EMPIRE, EMANCIPATION

As part of the Revolution's institutionalization, the nationalized energy industries were placed under the direction of the Ministry of Heavy Industry.

In the late 1950s, by contrast, three international oil majors controlled the Cuban oil sector. For Standard Oil of New Jersey (Esso), Texaco, and Royal Dutch Shell, Cuba was a small market, but clear of competition, the island provided an important outlet for their recent investments in high-priced Venezuelan crude. The exploitation of some of the world's largest proven oil resources started on a large scale in Venezuela in the 1920s. The oil frenzy reached a high in the 1950s when Venezuela's military ruler Marcos Pérez Jiménez worked with the majors to develop the country's oilfields.[84] In Cuba, *Revolución* wrote that this was a "fabulous business for the monopoly" since the multinationals could subsidize investments in Venezuela by exploiting the Cuban economy.[85] To Cuban nationalists, the arrangement gave more evidence to an international oil conspiracy, testifying to how peripheral Cuba served the world industrial core. National underdevelopment, and the conditions of possibility for development, was linked to a political economic compact that sustained the supply of oil to the island.

In the Caribbean, just as in other island regions, the absence of fossil fuel deposits and large rivers prompted governments to rely on oil imports.[86] In many other postcolonial states, modernization programs often involved the construction of large hydroelectric dams, such as Nasser's Aswan High Dam on the Nile and Stroessner's Itaipú Dam on the Paraná.[87] Experts in both Cuba and the Soviet Union rehearsed a narrative of Cuban resource scarcity. On more than one occasion Fidel Castro stressed how Cuba had no great rivers, no water resources that could be dammed, no coal deposits, and even a lack of wood. The dearth of wood was both an ecological and a political condition, he argued, since Cuba had been deforested during the colonial period, "by the foreign companies who used cedar, mahogany, whichever wood they used as firewood in the sugar *centrales*."[88] In another account, Boris Semevskiy, a renowned geologist from Leningrad University, remarked that "energy supply, considered in its most general form, is the problem of problems for each socialist country. . . . One of the problems in solving Cuba's energy problem is the inadequacy of local energy resources and the nation's dependence on oil imports."[89] The only non-oil-based energy sources of any significance were bagasse—the fiber that remains after cane has been milled—which was burned in the *centrales* during the harvest and a 28 MW hydroelectric dam which opened on the Hanabanilla River in 1963. The Cuban islands were known to lack resources, and the physical landscape thus legitimized, if not demanded, oil imports.

In February 1960, the revolutionary leadership invited Soviet First Deputy Premier Anastas Mikoyan to Havana to inaugurate a Soviet cultural exhibition. Facing mounting US hostility, Fidel Castro agreed to the invitation anticipating that Cuba, overtly or covertly, would have to rely on Soviet support in the future.[90] Castro and Mikoyan signed a trade agreement that provided Cuba with nine hundred thousand tons of oil in one year. According to a declaration in *Revolución*, Cuba would save twenty-four million dollars in hard currency.[91] The Cuban leader legitimized the agreement by invoking the sovereign nation's right to "trade extensively" and stressed that both nonaligned Argentina and US-aligned Brazil had signed similar agreements with the Soviet Union.[92] In Washington, the Eisenhower administration saw the agreement as a provocation and prompted Esso, Texaco, and Shell to "get tough" and refuse to process Soviet crude in their refineries in Havana and Santiago de Cuba.[93] In a broadcast, Fidel Castro declared that the resulting oil blockade was "the first great obstruction [*zancadilla*] against our Revolution."[94] By interrupting the flow of energy to the island, the United States disrupted the metabolism of the Cuban economy for political gain. As the conflict escalated, the Cuban government confiscated the refineries on June 29 and July 1, 1960. The United States then cut its sugar quota, which led the Cuban government to nationalize the refineries along with the Compañía Cubana de Electricidad, the Cuban Telephone Company, and thirty-six sugar mills on August 30, 1960.[95]

In the Soviet Union, the oil industry had become a key sector in the years following the Second World War. Oil production started in the area around the Caspian prior to the Bolshevik Revolution, before expanding to the Volga Basin in the 1940s and West Siberia in the 1960s. During Stalin's reign, the Soviet economy consumed almost all the oil it produced, but with increasing production, the Soviet Union became the world's leading oil-producing country with an output matched only by that of Saudi Arabia.[96] In the 1950s, Khrushchev embarked on an export-oriented strategy, putting oil and gas up for sale to both Soviet- and US-aligned countries. In the early 1960s, the Druzhba (Friendship) oil pipeline and the Bratsvo

(Brotherhood) gas pipeline linked the Soviet oilfields with Eastern and Central Europe.[97] When Mikoyan arrived in Cuba in 1960, a Soviet trade delegation also arrived in India, offering to supply approximately 50 percent of India's oil imports on a bartering basis and at a 20–25 percent discount. While Jawaharlal Nehru turned down the Soviet offer—the oil majors refused to refine Soviet oil in India too—the Soviet Union's deal with Cuba followed a trail of other agreements with countries including Argentina, Brazil, Ceylon (Sri Lanka), Egypt, and Ghana.[98]

The Soviet Union soon became the all-dominant source of Cuban oil supplies. In 1972, when Cuba prepared to adopt the SDPE, the country also joined the Soviet-dominated Council for Mutual Economic Assistance (CMEA). CMEA coordinated trade between the socialist states and acted as a counterpart to the Organization of Economic Cooperation and Development (OECD) in the West. As one of CMEA's underdeveloped countries—alongside Mongolia and later Vietnam—Cuba was offered oil in direct exchange for sugar with a settlement indexing the oil and sugar prices at a highly favorable rate for Cuba. Without this politically negotiated trade plan, both parties argued that the underdeveloped country would suffer from the structural economic inequality that otherwise existed in trade between agrarian and industrialized countries. Indeed, the membership in CMEA allowed Cuba to break free from the neocolonial world market, counteracting unequal exchange and enabling the country's development.[99] Cuba's CMEA membership tied the Cuban economy firmly to the Soviet-dominated economic sphere. By 1987, 87 percent of all foreign trade took place with CMEA countries and 72 percent with the Soviet Union alone.[100] In the case of oil, Cuba imported 16 percent of its supplies from the Soviet Union following the 1960 agreement (0.9 of 5.5 Mt); in 1987, Cuban oil imports peaked at 13.5 Mt with 89.3 percent (12.1 Mt) of the total supplies arriving from the Soviet Union.[101]

Cuban-Soviet trade was operationalized through an infrastructural system with transatlantic reach. Initially, Soviet tankers arrived in Cuba from Black Sea ports, such as Novorossiysk. In 1961, Ventspils in the Latvian SSR became the main hub of Soviet oil shipments westward, to western Europe and Scandinavia, and an extension of the Druzhba pipeline connected Ventspils with the oilfields in West Siberia. Owing to the greater proximity between Cuba and the Baltic compared to the Black Sea, one contemporary observer concluded that "great savings are . . . made from transporting crude from Ventspils to Cuba which gets most of her crude from the USSR."[102] In official Cuban and Soviet rhetoric, the territorialization of this transatlantic infrastructural space, enabling socialist development, testified to the solidarity and the fraternity existing between the Cuban and Soviet peoples.

The availability of oil in Cuba, like the socially uneven landscape of electrification prior to the Revolution, was a question of political economy. On the eve of revolution, this made Cuba's energy infrastructures a central battlefield in the struggle over the country's socioeconomic future. By sabotaging key nodes in Havana's electricity system and blockading oil supplies to the island, dissatisfied electricians, refinery managers, and even the US Navy disrupted the supply of low-entropy energy into the Cuban economy. Through sabotage, they exercised a tactical kind of power associated with energy use whose mode is immobility, stalling the metabolic process, and whose medium is infrastructure.[103] To the saboteurs' despair, the revolutionary government seized control over the energy industries. Soviet oil, traded in return for sugar, provided the material foundation for an economy configured to undermine the law of value. It also counteracted unequal exchange in trade between agrarian and industrialized countries. In a physical landscape shaped by a long history of colonization and extraction, the limits of political possibility were redefined in relation to oil and electricity and the terms of their distribution.

2 Electrification or Death

Southward, toward the mouth of the bay, the silhouette of a great dome broke the horizon. I was sitting in the shade of a ceiba watching an intense baseball game; a faded mural demanded "Socialism or death!" and two workers were busy loading cement bags onto a Soviet-era truck. Black smoke puffed from it. When the dome came into view across the bay of Cienfuegos in the 1980s, it was intended to be the Revolution's greatest achievement—a concrete-and-steel structure that testified to Cuba's social development. Alone, the first nuclear reactor at Juraguá would provide the country with more electricity than all prerevolutionary power plants together. But seeing the reactor shell three decades later, the abandoned construction site seemed to testify to something else. It spoke of a vision of social progress that had been intimately linked to a materially distinct infrastructure, an infrastructure again tied to what had been a particular political economy of energy supply and distribution. In the reactor dome, narrative, materiality, and political economy laced together into an energy infrastructure that would facilitate that socialism on the wall.

EXPANSION: "EVERYTHING REQUIRES ELECTRICITY!"

Shortly after coming to power, the revolutionary government set a target to double Cuba's electrical capacity. Between 1961 and 1965, the state would install 600 MW worth of generating power.[1] The target was set so that electrification would proceed, at least, at the same rate as industrial growth in general.[2] To make this possible, Che Guevara had traveled to the Soviet Union in October 1960, securing an agreement including imports of power plants and guarantees of technical assistance.[3] The nationalized utility soon also synchronized new power plants with the country's two main electricity networks. In February 1966, the thermoelectric plant "Máximo Gómez" opened in Mariel with a first 50 MW generator. Construction had taken four years to complete, and the workers had been forced to dig out an artificial lake to provide the steam boiler with water.[4] Management had also been pressed to set up a Communist Party cell to quell a group of workers forming a union. According to a local Party official, unionization was "a habit" lingering from the time the workers were employed by capitalist companies. Now that the workers owned the factories themselves, syndicalist tactics were at odds with the devotion to work necessary for the Revolution to succeed.[5] Hours after the "Máximo Gómez" came online, the "Antonio Maceo" opened in Renté, Santiago de Cuba, adding another 50 MW to the grid. Both plants were built with Soviet equipment, "from the generators to a system of chemical cleaning of the boilers," and Soviet technicians assisted the Cubans in the construction works.[6] Months later, the "Máximo Gómez's" capacity had increased to 200 MW and the "Antonio Maceo's" to 100 MW.

Like all state enterprises, the power plants were christened after martyrs of the anti-colonial struggle. This was a symbolic act inserting them into a revolutionary imaginary. With Gómez and Maceo, two heroic generals attested to the power plants' significance. Inaugurating the Renté plant, Minister of Industry Joel Domenech declared that "in each of these pieces of machinery, the sacrifice and blood of Santiago's best sons is present, the Moncada is present, the Granma, the Sierra, Frank País and Pepito Tey."[7] Tey and País had led the M-26-7's urban resistance in Santiago de Cuba and had been killed by Batista's forces in 1956 and 1957. In the revolutionary narrative, the attack on the Moncada Barracks, the landing of the Granma, and the campaign in the Sierra represented three key moments in the M-26-7's armed struggle. Three years later, the utility synchronized the "Diez de Octubre" (128 MW) in Nuevitas with the Occidental Electricity System, the name referring to the day when Carlos Manuel de Céspedes, a white plantation owner, freed his slaves and declared Cuba independent from Spain in 1868. Hence, this power plant was also symbolically inserted into a longer trajectory of national liberation. A few years after the Nuevitas power plant was inaugurated, a new thermoelectric plant in Cienfuegos with a 66 MW generator from Czechoslovakia was in turn named the "Carlos Manuel de Céspedes." The power plants both symbolically and physically enacted the revolutionary project.

In early 1969, Germán Wettstein, a Uruguayan geographer, visited the construction site of the Nuevitas plant. A staunch supporter of the Revolution, he was invited on a cultural mission to observe the Revolution's achievements.[8] The revolutionary government deemed the location of Nuevitas on the Camagüey north coast exemplary for an industrial center. As the seat of colonial administration, Havana had become a "parasitical city" that drained its hinterlands of resources and laid claim to industrial investments serving its white elite. The revolutionary government therefore decided to neglect Havana for the benefit of the deprived regions.[9] Alongside the ports of Cienfuegos and Santiago de Cuba, Nuevitas became a priority area for rapid industrial development. Wettstein's travel memoir shows how the symbolic and physical dimensions of the Nuevitas site blended into a visceral experience of the Revolution and its emerging infrastructure. Next to a factory for electrodes and barbed wire, and another for cement, Wettstein described a mosquito-infested construction site where 500,000 m³ of solid materials had been removed before the prefabricated Czechoslovak power plant was assembled. More than eighteen hundred technicians worked ten-hour shifts, twenty-four days straight, before a six-day break to visit family. To Wettstein, the work enthusiasm he witnessed merged with the sight of the plant, which was crested with an illuminated sign. He writes that "24 hours a day, every day, the thermoelectric [plant] is a large beehive, but above all at night, when the lights of the humanized landscape are reflected in the waters of the bay and at the top of its first generator the three letters CHE shine bright, the emotions run deep."[10] Again, materiality and narrative laced

Figure 4. The thermoelectric plant "Otto Parellada" in Tallapiedra, Havana. The baroque-style building on the right is the original, now defunct Tallapiedra plant. The striped chimney together with the oil cisterns in the foreground are part of the modern-day thermoelectric plant, which opened in 1973 boasting a 64 MW generator. Photo by Gustav Cederlöf.

together in the emerging energy infrastructure, this time evoking the ethos of Che Guevara.

While the government's overall economic strategy shifted from the guevarista ideals of the Revolutionary Offensive to the adoption of the SDPE in the 1970s, electrification and increased energy consumption remained primordial revolutionary tasks. In addition to new power plants, the nationalized utility installed new generators in already existing plants. In 1972, for example, a 60 MW unit from Czechoslovakia was fitted in Tallapiedra; in 1978, a 100 MW generator in Mariel, another 100 MW generator in Santiago de Cuba, and a 169 MW generator imported from Japan in Cienfuegos.[11] When it met for its first congress in 1975, the PCC decided to increase the national electrical capacity by 850–1,000 MW over the coming 1976–80 five-year plan.[12] At the end of the plan period, the Party reported that 1,069 MW had been installed, doubling the capacity since 1975.[13] Between 1958 and 1978, electrical output grew by 7.2 percent annually, essentially based in five large oil-fired power plants (Figure 4). Speaking in

Tallapiedra in 1972, Fidel Castro reasserted the importance of electrifi-cation to the Revolution without wasting words: "Everything requires electricity!"[14] While the prerevolutionary electricity systems had relied on smaller generators distributed in a multiplicity of independent grids, elec-tricity consumption under the Revolution increased most efficiently with generation concentrated in a few nationally interconnected places.

REDISTRIBUTION: "TO THE LAST CORNER OF THE ISLAND!"

Beside the imperative of increasing energy consumption, a notion of distributive energy justice shaped the geographical profile of the new electricity system. In 1953, Fidel Castro presented his manifesto in the so-called Moncada Program, and contrary to the efforts of the privately owned Compañía Cubana de Electricidad, he promised that the entire population would have access to electricity after the Revolution. Under a progressive government, the electric current would reach "to the last cor-ner of the Island."[15] Understandably, electricity generation in a few specif-ically designated areas was also only useful if these areas were connected to the country's dispersed households and industries. The two large and more than sixty smaller systems were to be fused into one territorially encompassing, nationally integrating infrastructure.

In February 1973, the Oriental and the Occidental Electricity Systems were interconnected with a 110 kilovolt (kV) cable between Nuevitas and Holguín. This established the Sistema Electroenergético Nacional—the National Electricity System—or SEN.[16] In line with the 1976–80 five-year plan, the utility then upgraded the grid with 220 kV lines to increase its capacity and reduce heat losses in transmission.[17] As part of the "in-stitutionalization" of the Revolution, the government also renamed the Compañía Cubana de Electricidad. With a twist on the new name Unión Eléctrica (UNE—Electrical Union) and the national grid, the company's slogan became Lo que nos UNE—"that which UNITES us." The unifying grid extended progressively over the island. In June 1986, the PCC's news-paper Granma announced that the municipality of Maisí in the province of Guantánamo, the last municipality without a connection, finally had been integrated into the system with a 33 kV line from Baracoa. The linesmen had vanquished rivers and mountains to draw the cable, the newspaper reported, and the connection now fulfilled a revolutionary objective set out in the Moncada Program.[18] By the mid-1980s, the SEN interconnected the entire nation as one revolutionary socio-spatial whole.

Several economic and social interests legitimized the spatiality of the Revolution's infrastructural form. First, Che Guevara argued that an infrastructure encompassing the entire country would provide the state with the means to effectively localize factories. With lines weaving through Cuba as "one immense zone," industrial production could take place without taking the availability of energy into account. If a grid connection was missing in advance of industrial development, the problem could be solved quickly "with cables that will run at the same rate as the investments."[19] Second, the unification of the many systems into one centralized grid would also rid the country of inefficient energy supplies. To Guevara, inefficient systems were smaller scale with isolated power plants; an idea resting on the notion that large steam boilers and turbines generated economies of scale. By integrating smaller power plants into a capacious multi-plant system, the energy system would become more stable and it would be possible to even out the supply across the country.[20] In Pinar del Río, the isolated 9 MW "Eliseo Caamaño" plant had served the city since 1929. In a first step toward integration, the plant was incorporated in a provincial network before it joined the SEN at Los Palacios and ultimately was decommissioned in the 1990s.[21]

These reasons for an interconnected grid notwithstanding, the most important reason was that a uniform infrastructure would integrate the underdeveloped countryside with the urban centers. Revolutionary energy justice was a matter of equality in distribution but also of social transformation. In the Moncada Program, Fidel Castro explained that 2.8 million people lacked access to electric light in rural and suburban areas.[22] Though Cuba had a relatively high rate of electrification compared to other Caribbean countries, only about half the population had access to electricity. As Lillian Guerra notes, "Cuba's 'development gap' was so great it appeared to reveal two different countries." Urban households accounted for 87 percent of the connections, while rural dwellings stood for 9.1 percent.[23] Beyond the urban-rural divide, the government explained the country's uneven electrical development in class-based terms. Cuba had been "a country where electricity was used to supply light to the most aristocratic *barrios* in the city,"¹ Guevara argued, "to supply lots of light to illuminated signs, for unnecessary expenses."[24] This injustice had developed because of the Compañía Cubana de Electricidad's profit motives, which only served its US-based owners. The Compañía "extends the lines as far as it is profitable," Fidel Castro charged, "and from there on they don't care if people have to live in darkness for the rest of their lives."[25]

The focus on rural electrification worked in tandem with the government's "anti-urban" agenda, focusing on the deprived regions at the expense of Havana. In Fidel Castro's words, the only way for the peasant to free himself from his misery before the Revolution had been to die, but the Revolution brought development to the countryside.[26] And yet, the government's reforms in areas ranging from land and housing to education, health, and electrification were not meant to improve the life of the peasant as such. Echoing the GOELRO plan, the goal was to transform the conditions of rural life to such an extent that the peasantry withered away as a social class. The peasantry would disappear when the productive forces developed, the entire country was electrified, and production was socialized into larger units. The Revolution would engender a radical spatial transformation of Cuba, improving the standards of living so that all Cubans became members of the urbanized workers' state.[27] Against this line of reasoning, Fidel Castro exhorted the peasants to move toward "superior forms of production" either by seeking employment in the state sector or by forming cooperatives eligible for subsidies and credits for mechanization.[28] When the productive forces developed in the countryside, Che Guevara concluded to the First Forum on Electrical Energy, the Revolution would be in the process of eliminating the unjust differences that characterized urban and rural life.[29]

INFRASTRUCTURE AS SCALAR STRATEGY

Despite the electric current's seemingly immanent ability to transform social life, electrification was only meant to transform a determined set of social practices. Prior to the Revolution, the creation of an urban consumer culture was seen as an enabling condition for electrification, focused as the utilities' strategies were on gendered household practices. The kitchen was a key space in which everyday practices could be transformed to drive a demand increase for electricity. In the socialist state, by contrast, it was not market demand but state priorities that guided investments. The government saw national industrialization as the most important contributing factor behind a demand increase, but even so, when the PCC rehearsed the Revolution's achievements in the electrical sector, it also reported on the number of electrical appliances the state had distributed to the population: between 1976 and 1980 there were 770,000 televisions; 420,000 refrigerators; 1,350,000 radios; 465,000 washing machines; more than 400,000 electric fans.[30] In contrast to the private companies making appliances available through the market, the state offered households a select range of appliances, fashioning the technological environment of everyday household life through central planning.

At first, a coupon system was set up for the distribution of domestic appliances, which soon was integrated into the state's rationing system. The government also set up the *emulación socialista* (socialist competition) in 1971 in which the CTC rewarded workers with household appliances for meeting production quotas and participating in political acts. As María Cabrera Arús argues, the distribution of consumer goods played an important part in the process of state formation since it consolidated the state bureaucracy through practice.[31] Largely, however, energy use in Cuba's kitchens, which remained a strongly gendered space, was not based on electricity but kerosene, LPG, and denatured alcohol; not "the electric stove, of course," Fidel Castro asserted, "because of its enormous electricity consumption."[32] The historical underdevelopment of the electrical industry, he argued, meant that the available infrastructure was unable to sustain the load added by electric stoves. Parallel to the electrification campaign, household consumption of kerosene and LPG instead increased in an almost linear fashion.[33] As a result of the government's selective allocation of resources, then, everyday household practices developed around the use of liquid fuels rather than electricity, which shaped the gendered experience of cooking.

The state not only distributed household appliances but also cooking fuels through the rationing system. This system was set up in 1962 based on the idea of the socialist state as a guarantor of distributional equality. As part of the system, food, fuels, and consumer goods were construed not as commodities with exchange value but as entitlements distributed for the population to *alimentarse* (to nourish themselves).[34] With a spatial metaphor, the socialist state can be pictured as an inverted funnel that draws produce in from across national space before it redistributes the same produce equitably among the population. The institutional mechanism to achieve the funnel effect with respect to food was the state company Acopio, to collect the harvest, and the Ministry of Interior Trade (MINCIN) to redistribute it through a system of ration shops (*bodegas*) and ration cards (*libretas*). Similarly, kerosene, LPG, and denatured alcohol entered the funnel under the aegis of the national oil company Cupet to be distributed by MINCIN in the bodegas. In a bodega in eastern Cuba in 1990, for example, the anthropologist Mona Rosendahl observed that kerosene was available in 3.5-gallon units (13 L) to cover the needs of one person, or 9 gallons (35 L) to cover for a family of five.[35]

The funnel effect reflected the universalization of property in the socialist state, echoing Che Guevara's ideal of the nation-state as an integrated factory. As Marisa Wilson argues, it was "a model that treats the entire nation as one socialist enterprise, whose ultimate aim is not profit (surplus value) but to ensure alimentary and other needs (social values) of the national community."[36] The rationing system thus concretized a political strategy in which infrastructure was envisaged to serve social interests on a particular scale.[37] The Revolution came into being through infrastructure as an integrating, national project with the aim of radical social transformation. Cubans received electricity, kerosene, and LPG not as marketed goods but as non-commodified entitlements. As use values, these particular energy forms metabolized social practices that signified convenient, revolutionary life, confirming Matthew Huber's assertion that use value "is not natural or fixed but rather only emerges out of particular historical geographies, lived practices, and meanings through which a commodity [or entitlement] comes to be imagined as *useful*."[38] Addressing the workers who had installed a new generator in Tallapiedra, Fidel Castro reaffirmed that in the socialist state, they would reap the benefits of their own work. In doing so, he also reaffirmed the spatial logic of the Revolution's infrastructural form:

All of you, those who have participated in this task, from the workers whose professional work is construction to the thousands and hundreds of thousands of voluntary hours that tens of thousands of citizens have dedicated to this industry, you know that you will be the direct beneficiaries of this plant when it is started up (APPLAUSE), that you yourselves will use this energy or that the economy will use it in its industries for you, or for the development of the country, which is a development for you and for your children.[39]

To fulfil its social ambitions, the revolutionary government reconfigured the geography of everyday energy use. The SEN enabled energy use to be sustained in one place by generating vast quantities of electricity in another. "Once fused in the same site, [generation and application were] cast to the two extremities of a line," as Michael Anusas and Tim Ingold write with reference to the conditions of modern electrified life.[40] This re-spatialization of energy use is similar to other socio-ecological transformations associated with modernization projects. In their appraisal of swidden and Green Revolution agriculture, Michael Dove and Daniel Kammen argue that "development" does not seem to occur "unless produce enters a system wider than the local one in which it was produced."[41] Looking at roads, Penny Harvey writes that when framed as a national project, a road brings a scale into being that renders local interest subservient to that of a "'public work' that may be located but is no longer local."[42] In the case of mining, Alf Hornborg sees a process of modernization as one that "abstracts, encompasses, and disempowers the local," and to him, the notion of modernity is therefore neither a historical period nor a state of technological development, but a system of power in which resources and risk are unequally distributed.[43] In Cuba, the construction of the SEN re-spatialized energy use across scales to enable just socialist development.

The production of socio-spatial order cannot be separated from a particular kind of knowledge co-constructing the physical infrastructure. In its statistical reports, UNE described the development of the electrical grid in thermodynamic and other physical terms. This ability to produce statistical knowledge was vital to the government's ability to demonstrate revolutionary progress. Materiality and narrative welded together so that the "metrics of investment index[ed] state commitment."[44] When national development was tantamount to increasing electricity consumption, new generators—with higher capacity—generating more electricity—indicated progress in precise terms of megawatts. When social justice implied that electricity should be distributed equitably over territory and population, new power lines—covering greater distances—interconnecting more people—signaled progress in exact measures of kilometers of lines and share of the population with connections. Fidel Castro's speeches frequently exhibit this conflation of materiality, measurement, and narrative. Incessantly, he drew on UNE reports to compare the current state of the grid with figures from before 1959 in order to showcase the Revolution's accomplishments. In 1978, for example, he inaugurated a new generator in Mariel and detailed the material characteristics of the SEN:

> Comrade workers of Mariel:
> If you don't get bored, I can give you a few interesting facts on the topic of electricity, which are very important. . . .
> In 1958, the Compañía Cubana de Electricidad generated, in the two existing electrical systems, 1,760,000 megawatts/hour. . . . That is to say, before the Revolution, in 1958, 1,760,000 megawatts/hour; last year, 6,557,000 megawatts/hour, which represents 3.7 times more than that generated in 1958.
> This increment equals an average annual growth rate of 7.2%.
> Regarding the number of consumers, counting all houses that were receiving electricity, centers, public services, there were in 1958, 722,000 centers of consumption; in 1977 we had 1,343,000 centers of consumption.
> In 1958 13,100 kilometers of lines existed in the country across all voltages; in 1977 we account for 36,200 kilometers of lines.

The physically and institutionally centralized grid thus made the Revolution quantifiable as a national undertaking, enabling Fidel to reaffirm the government's success and commitment: "Since the triumph of the Revolution extraordinary attention has been given to the electrical development of the country, as a fundamental premise for all our economic and social development."[45]

Enabling an account of revolutionary progress, the SEN's wires enhanced the state's territorial presence, symbolically and materially. The energy infrastructure incorporated people's homes and household appliances in the energy system's relational configuration, even as the state utility extended its bureaucratic presence with the energy system's wires, gatekeeping the interface between sockets and plugs. By extending an infrastructural system, Michael Mann argues that the state exercises its infrastructural power to "penetrate and centrally co-ordinate the activities of civil society through its own infrastructure," and thus, when Cubans switched on the light, they also interacted with the socialist state.[46] At the moment of energy use, Cubans created use value with the energy system, partaking in the socialist project and giving life to the socialist state, just as people in a capitalist society partake in that project when they interact with energy infrastructures under different political economic and symbolic circumstances. And yet the constitutive moment of human-infrastructure interaction indicates that the state, rather than being an actor with infrastructural means, in some sense *was* the infrastructure. Instead of "penetrating" a detached civil society, the extension of the SEN was a way of tying together, of coordinating energy practices and creating social relations. More than a means of politics, the SEN was a mode of politics—a politicized form of socio-ecological organization.[47]

As in Cuba, the socialist government in Vietnam framed national electrification as a project of infrastructural democratization. After Vietnam gained independence from France and was partitioned in 1954, the North Vietnamese were enrolled in the construction of an electricity network under the banners of *"Đảng là ánh sáng"* (The Party is the light) and *"Dòng điện không bao giờ tắt"* (The current never stops). During the war with the United States, described by some as an attempt at "forced demodernization," they then took part in its repair, literally and metaphorically working to keep the current flowing.[48] Christina Schwenkel shows how the act of participating in the electrification campaign created a sense of belonging to the postcolonial project. By electrifying the country, the Vietnamese were engaged in the revolution, actively taking part in it, and the memory of the campaign later created a mode of address by which they were interpellated as socialist subjects. The energy transition, then, was co-productive of a political imaginary of social, economic, and environmental change, which legitimized particular infrastructural form.[49]

Energy infrastructure in part constituted the socialist revolution as a lived political project. In Cuba, the SEN and the rationing system facilitated increased energy consumption while serving as socio-technical vehicles of energy redistribution. Mann writes of the state's infrastructural power, but the Cuban energy system speaks to a mode of power that not only operates within a politicized environment but that produces an environment itself, through the orchestration of energy flows.[50] The socialist state was a process in which energy systems provided a scale-based form for human life. The Revolution's infrastructural form had a narrative and an aesthetic dimension through which technology and energy use attained meaning as part of a larger social project. Conversely, energy use and technology, as objects of knowledge, in part constituted the meaning of that social project. The infrastructure also had a political economic dimension, as it orchestrated the distribution of energy across territory and population. Institutionally, the state utility maintained and administered the integrated nation-spanning energy system. Not least, the infrastructure had particular physical properties, shaping revolutionary practice.

AHORRO: SAVING ENERGY TO SALVAGE THE REVOLUTION

In 1986, 51,692 km of cables crisscrossed Cuba to interconnect the country as one socio-spatial unit.[51] Just as most transmission lines worldwide, the cables were made from aluminum with a steel core. After silver, copper, and gold, aluminum is the most conductive metal, but since aluminum is both lightweight and relatively inexpensive to produce, it has become a global transmission line standard. When electricity passes along a wire, the wire heats up and starts to sag, which is why aluminum lines are given a steel reinforcement. The benefits of aluminum notwithstanding, the newspaper *Granma* noted in 1986 that more than 17 percent of all electricity in the SEN's lines dissipated as heat in transmission.[52] Engineers know the underlying phenomenon as Joule's First Law: when a current passes along a wire, the wire's resistance generates heat and increases entropy in the energy system. In an alternating current system, where the current varies over time, the current also causes resistance to itself (known as reactance), which creates further heat losses. Out of every one hundred barrels of oil burned in the thermoelectric plants, therefore, the transmission infrastructure itself consumed seventeen barrels.

Beside energy losses in the SEN's wires, the thermoelectric plants could only convert a fraction of the energy stored in each barrel of fuel oil into electricity. The revolutionary strategy had been to increase the SEN's capacity with new installations almost exclusively. In the 1980s, there were even long-term plans to expand the transmission network with 500 kV cables, even as construction started on new thermoelectric plants in Santa Cruz del Norte, Matanzas, and Felton.[53] To achieve the highest possible capacity increases, the government refrained from replacing aging equipment. This meant that in the 1980s generators from the 1940s were still running at full speed with a much higher fuel consumption rate than the most modern machinery. In 1984, for example, two prerevolutionary units in Cienfuegos reportedly consumed 420 grams of fuel oil per kilowatthour (g/kWh), in comparison to the Czechoslovak generators installed in 1978 that consumed 217.5 g/kWh.[54] Fidel Castro lamented that the Cuban power plants on average could only convert 32 percent of the available energy potential into electricity, the remainder dissipating as heat.[55] On a different occasion, he explained that the old plants "are still around because we can't stop anything here."[56] No generator could be decommissioned before the Revolution had secured the electrical capacity necessary for the nation's development.[57] Economic growth was necessary for a successful transition to communism. Hence, the electricity infrastructure itself consumed oil by means of its materiality; a materiality that embodied and enabled Cuba's social future. Without a steady supply of oil, the infrastructure would become a jumble of mostly useless wire.

In the late 1970s, problems surfaced in the Soviet oil industry. Exploration and production teams in the core West Siberian region failed to meet their plan targets for the first time. With peaking output, production required deeper drilling and exploration had to move farther east where harder climatic conditions and longer transport distances were involved. This was a severe setback for the Soviet economy seeing that energy exports to the West generated about 75 percent of all Soviet hard currency income.[58] To forestall an escalating crisis, Leonid Brezhnev responded with heavy investments in the oil industry. In the Eleventh Five-Year Plan (1981–85), energy production accounted for two-thirds of all Soviet capital spending. Just like Cuba's investments in the SEN, however, the Soviet government prioritized new equipment at the expense of the increasingly inefficient machinery already in place. The marginal costs of production continued to increase.[59]

The gravity of the situation led Brezhnev to initiate a broader policy shift. As one of his first initiatives, Brezhnev's successor, Yuri Andropov, presented a full reform program in April 1983. In a radical move for the Soviet economy, modeled as it was for heavy industrialization, the reform aimed to free up oil for hard currency export by reducing domestic demand. To mark the reform's importance, Andropov presented it to the Central Committee of the Soviet Communist Party as "a GOELRO for today's conditions."[60] On the one hand, the reform comprised a program for fuel switching. Oil would be freed up in the electrical sector by replacing oil-fired thermoelectric plants with natural gas turbines and nuclear reactors. Already in 1980, the Soviet government had launched a major campaign to increase the use of natural gas in the electrical industry.[61] On the other hand, the reform encompassed a program for energy conservation. State enterprises would dispose of outdated machinery, and with improved metering, the Ministry of Electrical Power would work with price incentives to curb consumption.[62]

In Cuba, Soviet oil had become both an enabling factor and a condition for socialist progress. However, beyond satisfying domestic demand, official figures indicate that Cuba also benefited from its exchange of sugar for oil in other ways. Soviet crude was added to the Cuban export basket in the late 1970s. According to Fidel Castro, the Soviet Union formally agreed to Cuban re-exports in the 1981–85 trade plan.[63] The re-exports generated mounting income. In 1975, the Cuban state had earned 2.7 million pesos from exports of naphtha produced in the nationalized refineries.[64] In 1980, exports included Soviet crude and added up to 168.3 million pesos. In 1985, they reached 621.2 million. This represented 10 percent of the country's total export value, with oil overtaking nickel as the second leading export category (raw sugar accounted for 74 percent).[65] According to Jorge Pérez-López, however, the oil re-exports represented a whopping 40 percent of all hard currency revenue.[66]

In sum, the implications of the Cuban-Soviet trade agreement were threefold. First, it established an infrastructural space for socialist international trade. Soviet oil shipments to Cuba sustained the SEN and the Cuban economy with energy-potent hydrocarbons. Second, the hard currency earned from re-exports allowed the Cuban government to import goods from non-CMEA countries and to invest in social services. When

investments increased the demand for electricity, the state could easily expand its energy capacity with higher oil-import quotas. Third, the
abundance of oil permitted the government to use the oil it controlled
strategically in the international arena. Throughout the 1980s, Cuba re-
directed around ninety thousand tons of oil per year to Nicaragua, where
it also had military advisers and aid workers, in support of the Sandinista
Revolution.[67]

The prospects of short-term income from re-exports and long-term instability in the Soviet oil industry spurred changes in Cuban energy policy.
In June 1983, only three months after Andropov had presented his reform program, the government appointed a National Energy Commission
(CNE) to develop a new strategy. The commission reportedly consulted
more than two million workers in 47,925 workplaces and 44,650 trade-
union seminars.[68] A First National Forum on Energy then concluded the
process in December 1984. In the closing plenary, a banner extended over
the podium in Havana's Karl Marx Theater announcing that the country
would embark on a more rational, less oil-consuming path—not to "waste
even one kilowatt of electricity nor one gram of fuel or lubricant."[69] In a
nutshell, the national economy would maintain its levels of electrical and
industrial output, but closely resembling Andropov's reform, oil would be
freed up for re-export through fuel switching and energy conservation.
In government speeches, newspapers, and academic reports, the corner-
stone of the new strategy was condensed into one word: *ahorro*, or "energy
saving." *Ahorro* was essential, Fidel Castro argued, "because oil is not like
water that falls from the sky." The task of energy saving was so important it
even defined a revolutionary people: "We like to consider ourselves a revolutionary people, it appeals to us, we should even say that it satisfies our national pride. Oh!, but as long as we aren't a truly thrifty people [*un pueblo
realmente ahorrativo*], that knows to use every resource with wisdom and
responsibility, we cannot call ourselves an entirely revolutionary people!"[70]

The Energy Forum established three reform paths that would contribute to *ahorro*. The government had already taken measures along these
lines in recent years. The first was to reduce the consumption of electricity in households and industries in absolute terms, and the electricity tariff was used to incentivize energy conservation. In 1959, one of the first
orders of business had been to lower the tariff to just cover the costs of

production. With increasing consumption, the rates then decreased, from 6.5 centavos per kilowatt-hour (¢/kWh) for the first 40 kWh in consecutively cheaper price bands down to 2 ¢/kWh for consumption exceeding 200 kWh.[71] The aim of the falling rates was to incentivize energy consumption, inducing electrified development. However, twenty years later, *Granma* noted that the population's purchasing power had increased since the triumph of the Revolution and that people no longer prioritized energy conservation in the way they had when they needed to balance electricity consumption with other expenses. Consequently, it was rational for the government to increase the tariff. To disincentivize high consumption, the decreasing price bands were removed, establishing a flat 6.5 ¢/kWh regardless of consumption.[72] In the state sector, 150 work centers accounted for more than 25 percent of the country's total electricity consumption. For all enterprises consuming more than 50 kWh a month, the tariff almost doubled. During peak hours, it quadrupled.[73]

By modifying the tariff to "save" energy, the government reformulated the socio-ecological relationship between social development and energy use. Throughout the revolutionary period, increasing energy consumption had been heeded as a fundamental tenet of socialist modernization, but after the Energy Forum, this was only the case if energy was used "rationally" and with thrift—as it befitted a revolutionary people. Thriftiness, and hence the degree to which the people lived up to its revolutionary ethos, could be quantified in a form of scientific knowledge based on work output per energy input. Nevertheless, the new tariff did not alter the socialist state-citizen relationship established via the SEN. As almost the entire population was employed by the state, people received their salaries in exchange for work-contributions in the state economy. With the tokens of hard work their salaries represented, Cubans were entitled to electricity and other use values from the provisioning, redistributive state.[74] To help people reduce their electricity consumption, the Ministry of Heavy Industry recommended that doors and windows be kept "hermetically sealed" when using air conditioners. Electric stoves and water heaters—the two other "giants of electricity consumption"—ought to only be plugged in when foods were ready to be cooked.[75] *Granma* estimated that if households and enterprises continued to pay the same amounts for their electricity as before the reform, consumption would decrease by 700 GWh. This would equal a "saving" of two hundred thousand tons of oil.[76] In July 1981, eight months after the new tariff had been introduced, Fidel Castro announced that eighty thousand tons of oil had been "saved" in part thanks to the new tariff.[77]

The second reform path aimed to improve the efficiency of energy use in the state economy. CNE identified "100 measures in 10 areas" that would contribute to this task, ranging from better metering of consumption to guaranteeing complete airtightness in refrigerated spaces, to using the appropriate amount of lubricant oils in ball bearings, to introducing moral incentives for workers who contributed to reduced fuel consumption in their workplace.[78] Published with CNE's support, an even more elaborate guide expanded on these suggestions and drew on experiences from other countries that had embarked on similar energy-saving campaigns, most notably East Germany.[79] The press published mixed and inconclusive assessments of the initiatives that followed from these suggestions. After the first quarter of 1985, *Granma* reported that enterprises in the rural province of La Habana had met their production targets to 98.7 percent, but they had used only 86.3 percent of their assigned electricity, 89.2 percent of the diesel oil, 94.7 percent of the fuel oil, 91.2 percent of the gasoline, and 80.9 percent of the lubricant oils. This represented many thousand tons of oil "savings," the newspaper concluded, even as similar results also had been achieved in the capital city.[80] During the first twenty days of June 1985, by contrast, the state utility had generated 47 GWh more electricity than planned, exceeding its fuel-oil quota by over thirty thousand tons.[81] Eclipsing reports such as these, however, government discourse in the late 1980s mainly focused on the third reform path—that of fuel switching.

FUEL SWITCHING AND NUCLEAR MODERNITY

After three decades of Soviet-modeled industrialization attempts, the electrical industry was by far the country's largest oil consumer. The thermoelectric power plants used about a third of Cuba's total oil supplies, with the sugar industry following in lockstep.[82] In the *centrales*, fuel oil powered milling machinery, vacuum pans, and centrifuges while cane harvesters and tractors ran on diesel and lubricant oils in the sugar fields.

A habit of excessive oil use had developed in the sugar industry, Fidel Castro argued, because when the government had secured oil supplies in abundance, "the easiest formula was to open the tap."[83] In efforts to "save" oil, the Ministry of Sugar (MINAZ) began to substitute bagasse for fuel oil in its steam boilers. While the combustion of bagasse was a practice of old, deriving energy from eco-productive land, reintroduced bagasse use had a considerable effect during the 1982 *zafra*. Sugar mills in the provinces of Cienfuegos, Granma, and Guantánamo reportedly replaced all fuel oil with bagasse, and in a harvest of 8.2 Mt, the government announced that every ton of sugar had required 0.26 gallons (0.98 L) of oil in production. This was a substantial "saving" compared to 1981 when every ton had required 0.7 gallons (2.6 L).[84] In 1984, in comparison, Fidel Castro announced that four hundred thousand tons of oil had been "saved" in the production of raw sugar. Looking to the future, MINAZ planned on "saving" another four hundred thousand tons in the production of refined sugar, rum, and yeast.[85]

There were also efforts to harness energy sources other than cane biomass to replace imported oil. The government developed a long-term strategy to increase the national production of oil, which is discussed in chapter 4. Furthermore, since the mid-1970s, primary research had been carried out on renewable energy use in Cuban universities with a particular focus on solar-energy applications, bioclimatic architecture, and biogas production. This research gained attention in the 1984 Energy Forum alongside projects to develop the use of small hydroelectric plants and wind turbines in the agricultural sector.[86] However, from the government's point-of-view, the immediate prospects of renewables were limited due to the small amounts of oil they displaced. Fidel Castro argued that the development of renewable technologies, though important, would have to proceed independently of the new energy policy.[87] Instead, the central focus was on large-scale fuel switching in the SEN.

Correspondence between the Cuban and Soviet Academies of Science reveals that plans for a Cuban nuclear program emerged already in the mid-1960s. In the first instance, the Cuban Academy of Science set up a working group to develop research capacity in nuclear physics.[88] With the help of Michael Pentz, a British-South African physicist with Soviet connections who was employed at the European Organization of Nuclear

Research (CERN), the goal was to build a laboratory with three particle accelerators powered by van de Graaff generators. The lab would lay "the foundations for the peaceful application of atomic energy to the industrial and agricultural problems of Cuba."[89] In a meeting with Cuban president Osvaldo Dorticós, the Cuban academy director Antonio Núñez Jiménez and representatives of the Soviet nuclear industry promised that Cuba would find nuclear applications in areas as diverse as metallurgy, textiles, chemistry, geology, agriculture, medicine, and the electrical industry. The Soviet Union wished to sponsor Cuba with an experimental reactor.[90] Fidel Castro publicly emphasized the importance of nuclear energy for national development. Speaking in Playa Girón, he declared, "The energy of the future, the essential energy, the energy which humanity urgently must depend on in the future is nuclear energy."[91] In 1969, Núñez Jiménez finally inaugurated an Institute for Nuclear Physics with Fidel Castro, Raúl Castro, and Soviet Deputy Premier Vladimir Novikov in the front-row seats.[92]

Nuclear development had become a viable option for Third World countries in 1953 when US president Dwight Eisenhower launched his "Atoms for Peace" program. Western nuclear powers started exporting research reactors and radioisotopes to strategic allies, including but not limited to Israel, Iran, and India. As Gabrielle Hecht argues in relation to the Indian experience, "Indian scientists . . . saturated their atomic energy program with postcolonial significance, proclaiming it (and themselves) engines of modern nationhood."[93] A Cuban energy expert, whom I interviewed in 2015, also recalled that the Compañía Cubana de Electricidad had announced plans for a nuclear plant in Cuba in connection to Atoms for Peace. However, the Revolution cut its plans short. In the socialist sphere, the Soviet Union exported a reactor known as VVER-440 to its partners in CMEA. In the 1960s, a nuclear plant operated in Greifswald, East Germany, and by the 1970s, plants had been built in Kozloduy, Bulgaria; Bohunice, Czechoslovakia; and Paks, Hungary.[94] In 1976, as work also started on two reactors in Żarnowiec, Poland, the Soviet Union agreed to contribute to a nuclear program in Cuba.

The Cuban government envisioned a series of nuclear power plants to operate on the island. With four reactors each, there would be a first plant in the central province of Cienfuegos, a second in eastern Holguín, and

Figure 5. Model of the nuclear power plant "CEN Juraguá." In a propaganda film titled *CEN Juraguá: La obra del siglo (CEN Juraguá: The project of the century)*, which aired on Cuban television in the 1980s, the camera pans over a model of the nuclear plant then under construction. The speaker announces: "The construction of the first nuclear power plant requires the effort and participation of various organizations of government since the safe operation of it will represent a qualitative change in the national energy system." Nuclear energy represented the most advanced productive forces attainable in the world in the revolutionary narrative. Photo by Hvd69 / CC-BY-SA-4.0.

a third in western Pinar del Río. According to one forecast, nuclear energy would make up 15 percent of the SEN's total generating capacity by 1990 and 50 percent by 2000.[95] Work on a first reactor finally started in Juraguá outside Cienfuegos in 1983, with work on a second commencing next to the first two years later (Figure 5). To improve the capacity of the Juraguá nuclear plant, the government also planned on building a pumped hydroelectric station in the Escambray Mountains. While a thermoelectric power plant is easy to start and stop, a nuclear plant generates electricity twenty-four hours a day and is only shut down for annual maintenance.

This meant that there would be a surplus of electricity in the SEN outside peak hours. In a pumped hydroelectric station, or *hidroacumuladora* (hydroaccumulator) as it was referred to in Cuba, electric pumps move water upstream when there is an energy surplus and then generate electricity from the same water falling downstream during the demand peak. Together, the four reactors in Juraguá and the artificial dam would have a capacity of 2 GW.[96]

The government celebrated the Juraguá plant as a project of great historical and strategic importance. Cuban planners determined that each reactor would produce electricity equaling *ahorro* of six hundred thousand tons of oil per year. With all four reactors operational in Juraguá, the country would "save" 2.4 Mt annually.[97] If sold on the world market in the early 1980s, this represented approximately five hundred million dollars a year.[98] The Juraguá plant would also allow UNE to bring the most inefficient thermoelectric units offline and keep them as a backup reserve. The new thermoelectric plants in Santa Cruz del Norte, Matanzas, and Felton would be the last of their kind with a future all-out focus on nuclear energy.[99] Besides oil use in the thermoelectric plants, transports constituted an important part of the calculations. According to Fidel Castro, each megaton of oil arriving in Cuba required forty tankers to travel the forty thousand km from the Baltic, and 2.4 Mt of oil, by implication, required a fleet of close to one hundred fuel-oil-guzzling tankers. The uranium required for the same electrical output could be transported in a single train-car-sized ship.[100]

On the operational side, the nuclear program aimed to make the Cuban energy system independent of international relations. Czechoslovakia had notably been able to develop an integrated domestic nuclear industry in the early 1980s.[101] In the only substantial study that exists of the Cuban nuclear program, Jonathan Benjamin-Alvarado argues that Cuba could choose between a "turnkey" plant, operated by Soviet technicians, or to create a standalone nuclear industry.[102] The government appointed the country's leading nuclear physicist, Fidel Castro Díaz-Balart, who was the Cuban leader's eldest son, to coordinate the development of the latter. The Cuban Atomic Energy Commission (CEAC) trained Cuban nuclear physicists and radiochemists aided by the Soviet Union, the International Atomic Energy Agency (IAEA), and the United Nations Development Program (UNDP).

By 1992, around thirteen hundred Cuban specialists had graduated from universities in the Soviet Union and the Juraguá Electronuclear Polytechnic Center.[103] Next to the Juraguá plant, a purpose-built residential area—Ciudad Nuclear (Nuclear City)—was built to accommodate workers and technicians, showcasing the Revolution's ability to urbanize the entire country alongside its emerging nuclear infrastructure.[104]

Benjamin-Alvarado argues that the Cuban government opted for a standalone capacity since this was "the most flattering [option] for the political ambitions of the Cuban leadership."[105] For a postcolonial government, a nuclear program was indeed a prestige project, and the Juraguá plant turned into a national rallying object during the 1980s. At the same time, the plant became a rallying object in the United States. US representatives argued that Cuba was building a "Cuban Chernobyl" only ninety nautical miles south of the United States where a "major incident would create a radioactive cloud capable of creating serious ecological damage as far north as Tampa, Florida."[106] Fidel Castro Díaz-Balart instead insisted on the reactors' safety. The VVER-440 reactor, cooled and moderated with pressurized water, had a completely different design from Chernobyl's graphite-moderated reactors—a design, in fact that was originally based on Westinghouse technology. Eighteen reactors of a similar type were already operating in the United States. Like all sovereign nations, Cuba was entitled to nuclear energy, Castro Díaz-Balart argued: "The nuclear option not only constitutes an unquestionable necessity for the country: it is also a right."[107]

THE DIALECTICS OF DEVELOPMENT

More than prestige, the nuclear program was an attempt to "save" oil and reduce Cuba's dependence on the ailing Soviet oil industry. Nuclear capacity would reconfigure the political economic relations sustaining the flow of energy to Cuba. However, while nuclear energy would lessen Cuba's dependence on the Soviet Union, it would leave the spatiality of the SEN intact. In fact, the reactors would reinforce the Revolution's infrastructural form. A nuclear power plant expanded the SEN's generating capacity in one great leap. In a nuclear reactor, generating capacity would be based in an even more highly concentrated infrastructural node, representing even more complex productive forces and greater economies of scale. In the revolutionary narrative, nuclear technology signified the most advanced phase in the construction of the country's techno-material base. The reactors under construction in Juraguá cemented the structure of the SEN and the socialist state.

At the same time, the transition from oil to nuclear energy was intrinsic to the government's conception of history. In the 1960s, the Cuban leadership argued that oil dependence was necessary for the country to develop due to Cuba's lack of endogenous energy resources. In the early 1970s, Fidel Castro even noted that the country still was too underdeveloped to accommodate a nuclear plant. To be economically viable, the smallest reactor that could possibly be built would have a capacity of at least 300–400 MW. Such a plant would contribute more than one-quarter of Cuba's total electrical supply, and this would have fateful consequences when the plant came offline: "The entire country would come to a halt the day the units entered maintenance," Fidel Castro argued.[108] Put differently, the country had not yet reached a sufficient level of social development—consuming enough electricity—to advance to a more complex techno-material base. When construction started in Juraguá in 1983, the country had therefore developed to the extent that the plant could go offline without disrupting the metabolism of the Cuban economy. Oil dependence had generated electricity-consuming development to the point where oil was a steppingstone to a more advanced, nuclear techno-material base.

The Juraguá plant would soon make this socio-energetic dialectic swing back. First, nuclear energy would "save" the state substantial amounts of oil and bring hard currency in from re-exports. The revenue would allow the government to invest in more electrical appliances, more industrial production, and improved social services—all further increasing the consumption of electricity. A nuclear energy system would also enable new electrified social practices altogether. With oil-based generation, the state had distributed one set of electrical appliances to the population. But electricity was still unsuitable for cooking, as "inefficient" electric hotplates demanded too much electricity from the SEN. With nuclear power, Fidel Castro declared that the Revolution would modernize the country all the way into its kitchens: "When those nuclear stations are up and running . . .

there won't just be TVs, radios, refrigerators, fans, washing machines, irons, etcetera, but there will also be electric stoves. I explain this while I'm at it, so that you understand the importance of this type of industry that we have established in a place like Cienfuegos."[109] In the socialist conception of historical progress, Cuba moved from underdeveloped to developed, from "lacking" to "having," from exploitation to communism. This historical trajectory went hand in hand with the development of electricity infrastructure and the transition between biophysical energy regimes, from an organic to an oil-based to a nuclear electric economy. The relationship between society and energy developed with the productive forces, and the splitting of atoms would advance the Cuban Revolution into the twenty-first century.

3   Blackout

In 1990, the two nuclear reactors at Juraguá were still under construction. In their stead, Cuba's thermoelectric power plants fed electricity into the SEN and, hence, into households and workplaces. From Mariel to Santiago de Cuba, UNE generated 13.25 TWh of electricity, which made use of 3.34 Mt of Soviet fuel oil.[1] Years later, in 2015, I stayed long term with Anabel in Havana. She still had a "БК-2500" (BK-2500) air conditioner and a Soviet-made freezer in her home that she had put to use with that electricity twenty-five years earlier. In Pinar del Río, my neighbors José and Beatriz still relied on an "Аурика" (Aurika) washing machine and an "Орбита" (Orbita) electric fan. Electricity, together with kerosene and LPG, enabled Cubans to go about their daily household activities. These were energy sources distributed not as commodities but as entitlements to the socialist citizen. "*Antes* [before]," a transport worker called Manolito recalled with reference to Soviet times, "a liter of gasoline cost you 60 *centavos cubanos*. Now, look, it costs 1.20 CUC! [0.60 CUP = 0.02 CUC]. But the Russians gave it away—the fuel—and Fidel gave it to us and to other countries. A Russian tanker would come with gasoline and Fidel sent it off to Nicaragua. Such were the times. There was no economy."[2] With hindsight, and in what often seemed like a romantic light, energy use prior to the 1990s signified convenient, if careless, everyday life to many I spoke with all those years later.[3]

In the early 1990s, the infrastructural state started to unravel. At a time of rapid geopolitical change, Cubans were forced to enact an energy transition "under the shadow of emergency and exception." They had to break with oil less by political intent than, as Donald Kingsbury goes on to write, the need for "reactive decarbonization" in a context of acute oil shortage.[4] Nevertheless, when everyday practices changed to adhere to the new low-intensive, low-carbon situation, uneven local experiences reflected political priorities and social relations acting on scales far exceeding that of the energy user's experience. These were political priorities associated with a particular form of socio-ecological order, and while partly a technical matter, the reactive energy transition was above all a political and a cultural process negotiated in government offices, households, and workplaces.

"DEGROWTH" IN THE SPECIAL PERIOD

The energy crisis was particularly felt in the electricity sector. On Thursday May 28, 1993, Pinar del Río's residents could read in the provincial newspaper *Guerrillero* that "the blackouts this week, outside of the scheduling, occurred due to maintenance in the Thermoelectric Power Plant Carlos Manuel de Céspedes, in Cienfuegos, and a lack of fuel in its counterpart Diez de Octubre, in Nuevitas." The blackout timetable for the upcoming week then followed for each of the province's five distribution networks. The newspaper also informed its readers that "the scheduling of next week's blackouts could be considerably affected if conditioning works at the thermoelectric [plant] Antonio Guiteras, in Matanzas, would be carried out."[5] Every Thursday, statements like these were printed in *Guerrillero* while UNE, at the flick of a switch, put entire neighborhoods off grid. The shortage of electricity left buildings without running water when the utility's electric pumps stopped. There was no ventilation and no electric light. "After dark," Mona Rosendahl recorded in an ethnography from the time, "no housework could be done, no books or papers read, no television watched, and no meetings held. People often just stood or sat around in their yards smoking and chatting with their neighbors."[6]

From 1991 the situation worsened until its peak in 1993–94. Along-side the blackouts, the supplies of kerosene diminished, which about 75 percent of the population used for cooking.[7] In Pinar del Río, *Guer-rillero* announced that the plan for household fuel distribution would only be completed to 55 percent in 1993. The greatest problems were in the supply of LPG, which by June had been delayed for five months.[8] Initially, the state held reserves of cooking fuels together with food and medicines and distributed them through the rationing system. As the stocks were used up, the crisis intensified. The food situation was acute. The peso was inflated to insignificance, and soap sold at a third of the average monthly income. To make things worse, the "Storm of the Century" hit Cuba in the morning hours of March 13, 1993, months out of the hurricane season. "Those who lived through that night have never for-gotten it," Luis Ramos Guadalupe writes in his book on Cuban disasters; "and to recall it, they unwittingly refer to the images of the *Book of Reve-lations*."[9] Fidel Castro reported on the damages to the electricity system in the National Assembly, noting that it had rained almost 500 mm in twenty-four hours in one area of Pinar del Río: "Heaps of electricity posts were torn down, the electricity lines cut, transformers destroyed or dam-aged, the supplies of gas in Havana interrupted as a result of the lack of electricity; the water supplies, too[,] for the same reason. We have seen all these services affected in eight provinces."[10] The damages were esti-mated to be one billion US dollars overall, 20 percent of which occurred in the agricultural sector.[11]

In a strictly material, thermodynamic sense, the Cuban economy de-grew in the early 1990s. The quantities of energy and materials delivered to and degraded in the economic process declined fast. However, it would do a disservice to the argument for degrowth to characterize it simply in this way. It would also misrepresent the experiences of most Cubans. More than a material transformation, the degrowth goal is to create an econ-omy based on social and environmental values transcending the growth imperative: like the modernist socialism that had developed in Cuba, it is an anti-capitalist critique, but one that in some ways is even more radical than the case for socialism. As the anthropologist Susan Paulson argues, "Ideals of degrowth call us to shift value and desire away from produc-tivist achievements and consumption-based identities toward visions of good life variously characterized by health, harmony, pleasure and vitality among humans and ecosystems."[12]

The Soviet Union's collapse was a geopolitical shock that brought the Cuban economy into recession rather than a deliberate degrowth trajectory. Most Cubans were forced into energy poverty and were compelled to find solutions to the livelihood challenges facing them in the enforced low-carbon environment. And yet, to envision an intentional degrowth future, it is important to understand these challenges, just as it is to examine the socio-ecological implications of the solutions found to them. As Iris Borowy shows, the special period had certain positive health benefits even as the Cuban public health system was severely impaired. The forced changes to travel and food consumption patterns led to increased levels of physical activity and decreased levels of obesity, and Borowy concludes that "several years of living a life of economic decline and changed lifestyles left people similarly healthy or healthier than before."[13] For Sébastien Boillat et al., it is Cuba's shift to agroecology that signals "a real-life experience of degrowth," seeing that the already high levels of education and public health could be maintained despite the transition to low-impact agriculture.[14] Globally, an involuntary contraction of economic activity may be a politically more likely scenario than intentional degrowth, as it occurred in Cuba.[15] This again makes the implications of Cuba's experience of reactive decarbonization significant to study in some detail.

INEQUALITY AND THE POLITICS OF SCALE

For many in Pinar del Río and Havana, experiences of the special period cut deep through memory. One evening, I went to José and Beatriz's house to play chess. For José, the special period had been a continuous struggle for food and income; a time in his life that could hardly be reduced to daily blackouts but in which the shortage of energy in the centralized energy systems was a constant reality. Gradually during our conversation, I realized that José and Beatriz reacted differently to their shared memories. Confronted with some memories, Beatriz sighed intensely and trembled; "*No e' facil, no e' facil, no e' facil* [it's not easy]," she repeated below her breath. When José recalled how the state had distributed charcoal to households, she interposed: "To those who didn't have gas hobs" and closed her eyes. The special period is often spoken of as a collective experience, yet everyday life was strongly shaped by differentiated social factors. Alejandro de la Fuente has shown how the crisis augmented racialized differences, but experiences also varied across a gendered gradient.[16]

Julieta was only in her teens during the special period. She went to school close to her grandmother's house, and as she had a hard time eating the food offered in the school canteen, she was allowed to go home to have lunch. Julieta explained the difficulty of life in the special period, but in doing so, she implicitly also described the gendered geography of energy use in the household: Julieta's mother and grandmother cooked over a charcoal fire before serving Julieta, her father, and grandfather at the dining table. The women then ate whatever was left over in the kitchen. Julieta used to join them in this female space—"as children do," she said. "You know, I remember I asked my *abuela* [grandmother] why they were only eating chicken bones. 'Because we like them,' she told me. . . . But what do you tell a child?" Other women recalled experiences of cooking with charcoal or wood while describing their culinary inventions. "Have you heard about *el picadillo de la cá'cara del plátano*?" Rosa asked. "This was a common thing to do. People took the peel from bananas and minced it so that it looked like ground meat, and then prepared it just like *picadillo*—with onions and garlic. Or we'd eat 'soup,' which was basically *malanga* [a starchy tuber] in the water it had been boiled in."

José recollected the distribution of charcoal as a mundane memory, but it was clear that it had a more profound, embodied meaning to Beatriz. Beatriz remembered days in the backyard with her mother and sister-in-law, tending a charcoal fire. The homemade stove was long gone, but the surface where it stood was still there. The blackouts and the lack of cooking fuels had meant years of cooking with charcoal, a freezer and a refrigerator for long periods without cold, and a washing machine without whirl. The appliances she used in her gendered household practices required certain forms of energy input, and the Revolution's oil-dependent energy infrastructures co-constructed the techno-energetic environment with which she interacted. Beatriz's contact with the energy system thus enabled her housework and engendered the sociocultural significance of it. While reaffirming that use value only emerges in cultural context, a particular energy form can in this regard be seen as "an ingredient of the social practices and complexes of practice of which societies are composed," as Elizabeth Shove and Gordon Walker argue in their work on energy demand. Such a focus on praxis, however, must be sensitive to the wider power relations that shape the possibility of situated action and the "ongoing reproduction of practice."[17]

Several ethnographic studies demonstrate how housework is regularly sexualized as a non-masculine activity in Cuban public discourse.[18] This is also evident from my own experiences. One morning, for example, a colleague at the university came to work, as he always did, in a newly ironed shirt. I noticed nothing out of the ordinary and asked how his wife was. She had been ill with a high fever for over a week. He went on to greet another colleague who immediately smirked, "I didn't know that you could use an iron!" Our colleague blushed, saying that his wife was still in bed and that necessity knew no law: contrary to all expectations, and despite him being a trained mechanical engineer, he was learning how to iron. While the Revolution's social programs granted women economic and legal independence, the revolutionary woman was construed not only as a worker but also as a mother responsible for her family and by extension the nation.[19] In this context, Mélanie Josée Davidson and Catherine Krull note that it was women who were primarily responsible for acquiring alternative household fuels in the special period. It was women "who stood in long queues for food, cared for children and the elderly, volunteered for community work, strategized about making ends meet, and ministered to sick family at home or in the hospital."[20]

In the first half of the twentieth century, as we saw in chapter 1, investments in electrification were closely connected with the household as a space in which a demand for electricity could be generated. The creation of a gendered urban market for electrical appliances together with the production of the household as a gendered space made investments in infrastructure profitable. After the Revolution, by contrast, state planners were responsible for matching demand with available capacity even as housework remained a domain predominantly reserved for women. With the partial electrification of Cuban households in the geopolitical context of easy oil access, many of the work tasks assigned to women became less time-consuming and simpler, as the SEN enabled the use of refrigerators and other electrical appliances. During the special period, in comparison, women's relations to the energy system, within the gendered setting, rendered them experiences of a kind that most men did not have. Men and women did not share the same spaces for human-infrastructure interaction.

While Cubans experienced the scheduled blackouts and the fuel short-ages in situated social relations at the household scale, their experiences were also shaped by processes acting on a far larger scale than that of do-mestic energy use. At a national level, the government worked proactively to even out the consequences of the energy crisis. Like all kinds of human-environment interactions, situated practices of energy use should be ex-plained by moving analytically outward in space and backward in time.[21] From the government's perspective, the scheduling of blackouts reflected a political economic rationale of distributive energy justice, just as the design of the SEN before the crisis had reflected a state-socialist logic. The sched-uling of power cuts echoed the assumption that electricity should be evenly distributed across the population. Each week, the provincial branches of UNE received a set quota from the Ministry of Heavy Industry specify-ing how many megawatt-hours they had to save by cutting the electricity supply. Odalys, an UNE employee in Pinar del Río, worked with the time-tabling of blackouts in the province of Pinar del Río. "For each circuit we had to decide how long there would be a blackout," she explained. "You know, of course we tried to shut it down as little as possible at dusk when people really need the *corriente* [electric current]. But it was hard to do."

When the crisis was still in its infancy, Fidel Castro addressed the Com-mittees for the Defense of the Revolution (CDR) and argued that only a socialist system could confront the challenges in a rational manner. In a capitalist economy, the population's unequal purchasing power would be used to regulate the energy supply, distributing the available electric-ity by creating greater social inequalities. "In a capitalist system," he said, "what they would have done with the electricity at that moment, would be to double the price or triple it and not talk more about the issue. Every-one with little income, the poorest would be left without electricity and would be cut off." Under socialism, by contrast, the hardship would be borne equally by everyone through central planning: "Only a socialist re-gime . . . is capable of rationally addressing the supply of electricity and not solving the problem with prices."[22] Still ten years later, the magazine *Bohemia* made the same argument. In this instance, the socialist state had been forced to spend five hundred million dollars above the annual plan on oil imports following an international price spike. The magazine pointed out that "the electricity service to the population, zealously protected, was not made more expensive, but some economic activities like tourism and other industries, had to pay a higher price for the energy they demanded."[23] Irrespective of the levels of electricity it transmitted, the SEN was an infrastructure that assured the equitable distribution of energy as a non-commodity. The energy system warranted social justice by maintaining a tariff below the cost of production but with blackouts evenly spread across the network as a consequence of that decision.

Despite the rigorous planning efforts, the rationing of electricity did not always run as smoothly as the schedules suggest. José recalled how the blackouts would come and go, the exactness of the timetables notwithstanding. There were no certainties. At times, the reasons for extensive unplanned power cuts would be offered in *Guerrillero* and other newspapers, which typically involved unscheduled repair works, a lack of fuel, or load-balancing issues in a section of the grid.[24] From her perspective as an UNE employee, Odalys emphasized that power plants sometimes also unexpectedly could come online, "and then we would switch on a circuit that had been without electricity a little longer the day before." The government sought to mitigate the effects of the crisis on the provincial utility scale—the scale at which the socialist concept of energy justice worked in the special period—but the consequences of the energy shortages were experienced in the gendered context of everyday life, where the oven heated up, or not.

THE GEOPOLITICS OF A SCHEDULED BLACKOUT

While the rationing of electricity echoed a core revolutionary principle of distributive justice, the decreasing levels of energy consumption undermined another principle: the Leninist conviction that energy consumption must increase if a country is to transition to communism. In a direct sense, the challenge was the absolute lack of oil. Between 1960 and 1989,

Cuba's imports of Soviet oil products had increased almost sixteenfold, from 0.9 to 13.3 Mt.[25] When the Soviet Union collapsed, Fidel Castro explained that in 1959 1 ton of sugar had given Cuba between 7 and 8 tons of oil at world-market prices. When Cuba joined CMEA in the 1970s, and the oil and sugar prices were pegged to each other, the trade quotas were kept at a comparable level in successive five-year trade plans. Speaking in Santiago de Cuba in 1991, Fidel reported that the trade plans had given Cuba 1 ton of oil in return for 5 or 6 tons of sugar. In Cienfuegos a year later, he gave the equivalent of 1 for 7.5 tons.[26] "These are the famous 'subsidies' they talked so much about in the West," he told the Fourth Congress of the PCC.[27] Indeed, that the Soviet Union subsidized Cuban sugar is how most observers outside Cuba have characterized the trade relation in the aftermath of its collapse.[28] In the socialist narrative, however, the exchanges were regarded not as subsidies—a detraction from a rational market price—but as a just response to the reality expressed in the Prebisch-Singer thesis. The "distorted" prices were legitimate as they counteracted unequal exchange, and within the socialist political economic paradigm, the trade quotas were rational and fair.[29]

In his opening speech to the Fourth Party Congress, Fidel Castro explained the unjust economic mechanism underlying unequal exchange in the world system: The "industrial products that the developed countries sell become more and more expensive while the products from the developing countries, the countries of the Third World, maintain the same price or [have prices that] tend to diminish," he argued.[30] In Arghiri Emmanuel's seminal formulation, unequal exchange resulted from the industrialized countries' ability to import a surplus of labor-time embodied in commodities due to international wage differences. The continuous transfer of surplus value, in the Marxian definition, from periphery to core, led to globally uneven development.[31] Fidel argued that Cuba had experienced the effects of unequal exchange in the early years of trade with the Soviet Union when the two countries traded at world-market prices. However, when the two countries negotiated trade plans in CMEA, it ensured that the relative prices of products from industrialized and underdeveloped countries counteracted their built-in unequal exchange.[32] Inaugurating a cement factory in Cienfuegos, which had been imported from East Germany, Fidel Castro argued that trade with the already industrialized socialist countries made the Cuban economy "less dependent on the Western markets, on the highs and lows and on the crisis of this market; it makes it much less dependent on the unequal exchange that we have with the capitalist world."[33] In 1990, when he looked back on the past thirty years of socialist cooperation, he argued that the equal-exchange agreement they enjoyed with the socialist bloc had enabled Cuban development: "On these grounds we went on developing our country. . . . On these grounds, which were very fair . . . until our levels of development would be similar to the levels of industrial development of those [developed] countries . . . on these grounds, our industries were built, our agriculture was developed and mechanized; on these grounds the country was electrified."[34]

However, in late 1990, when the five-year trade plan from 1986 was coming to an end, no negotiations had taken place for a new agreement. The relations between Havana and Moscow had soured over policy differences. In the early 1980s, Cuba was caught up in the debt crisis that hit Latin American countries widely. Although Cuba traded mostly with CMEA, significant exchanges also took place with Western countries in hard currency, drawing on income from oil re-exports in particular. In 1986, Cuba re-exported 3 Mt of Soviet oil, but instead of the 621.2 million pesos earned from re-exports in 1985, the government found a gaping hole in its coffers when revenue fell to 269.1 million pesos following a slump in the oil market.[35] Cuba's debt-to-export ratio soared, and as Marifeli Pérez-Stable argues, the Soviet-inspired economic system "had reached a crossroads: broader application of market mechanisms or retrenchment."[36] Faced with this impasse, Fidel Castro declared that the country would go through a national Rectification of Errors and Negative Tendencies. The relative openness of the SDPE, with free farmers' markets and legal self-employment, would be replaced with a renewed emphasis on guevarista ideals and central planning spearheaded by the PCC.

In the Soviet Union, Mikhail Gorbachev instead embarked on perestroika—his program for economic restructuring—notably seeking to mitigate the problems facing the oil industry. Influential voices in the new Soviet leadership also questioned the equal-exchange agreement existing with Cuba, arguing that Soviet foreign relations had to become more mutually beneficial. Reminiscent of the contrasting perspectives put forward during the Great Debate, Fidel Castro argued that perestroika, which introduced further market mechanisms in the Soviet economy, was bound to bring the Soviet Union back to capitalism. Under Rectification, Cuba would instead reassert core socialist principles.[37] Despite their differences, Gorbachev made a state visit to Cuba in April 1989 to mark the thirtieth anniversary of the Cuban Revolution. Short of a trade deal, the two leaders signed a Treaty of Friendship and Cooperation that would last twenty-five years. However, it was only a statement of intention, and it was not until December 29, 1990, on the cusp of the new year, that a one-year trade plan was finally agreed.[38]

Reflecting the frosty relations between the once so close partners, Cuba's oil import quota shrank from 13.3 Mt in 1989 to 10 Mt in 1990. This meant that Cuba increasingly had to rely on the international markets for its oil supplies, markets on which the prices incidentally were soaring following Iraq's invasion of Kuwait. At the same time, Cuba had to rely on significantly lower international sugar prices for income. When Russia took over the Soviet Union's international agreements in 1991, Cuban-Russian trade remained focused on oil and sugar but now at world-market prices. For Cuba, this meant that 1 ton of sugar gave 1.4 tons of oil and that, by 1993, Cuban crude oil imports had dropped from 13.3 to 5.5 Mt.[39]

Publicly, Fidel Castro explained the situation as one inextricably linked to the Cuban people's toil in the sugar fields. The crisis could not be separated from the extractive economy and the long history of colonial exploitation in the Caribbean. Without socialist trade, the supply of energy into Cuba's industrial infrastructures was at the mercy of the capitalist sugar markets, the vicissitudes of a successful *zafra*, and the injustices of unequal exchange. At a mass rally in Cienfuegos, Fidel Castro reasoned, "This means that each thermal power plant, or each locomotive, or each truck, each vehicle is consuming sugar; it is like instead of pouring in tons of fuel, we would pour in sugar." As a question of sugar, the population would know the implications of the energy crisis not only in a rational, economic sense, but also emotionally, through bodily experience. Fidel related the image of the trucks running on sugar to the crowd's collective experiences of working in the cane fields: "You know what it costs to produce sugar: to sow and grow the cane, to cut it, transport it, to process the sugar, to store it and export it; see [for yourselves] if the problems

Figure 6. Diesel-powered sugar cane harvesting machinery. The first mechanical cane harvesters that could cut, clean, and load cane in one go were imported in time for the Ten Million Ton Harvest in 1970. During the 1972 *zafra*, 351 cane combines were in operation, but it was the introduction of the Cuban-manufactured KTP harvester—pictured above—that enabled the large-scale mechanization of the *zafra*. In the mid-1980s, about six hundred KTP harvesters were manufactured annually, and in 1990, there were more than four thousand units on the island. *Source:* Pollitt and Hagelberg, "The Cuban Sugar Economy in the Soviet Era and after," 553–54. Photo by iStock.com/vanbeets.

are severe or not" (Figure 6).[40] In the government's analysis, the oil shortages were a product of the Cuban islands' peripheral place in the neocolonial world economy: development was a question of geopolitical relations organizing asymmetrical energy flows. As Alf Hornborg has argued with reference to the experiences of many post-Soviet farmers, "When there is no longer any diesel in the tractor, it is just an assemblage of scrap metal. Again, what ultimately keep the machines running are global terms of trade."[41]

In a wider perspective, Cuba was far from the only country affected by the Soviet collapse in terms of energy use. As noted, the Soviet government developed extensive oil and gas pipeline networks in the 1960s, delivering energy across the Soviet republics but also into Eastern and Western Europe. With the dissolution of the political union, new international borders fragmented the once integrated infrastructural systems. For example, when Lithuania declared independence in 1990, Moscow responded by closing the gas taps in retaliation, just as it did in the Russian-Ukrainian conflicts several years later. In the Caucasus, Armenia, Azerbaijan, and Georgia deliberately disrupted the supply of energy to each other as they entered into conflict. Furthermore, the rapid commodification of energy in the transition from socialism to capitalism caused interruption in a more indirect sense. Oil and gas deliveries in the Soviet Union were determined in five-year plans, but when many former Soviet republics had to pay for energy imports from independent Russia and Turkmenistan—the major Soviet successor state producing natural gas in Central Asia—they accrued big debt leading to suspended deliveries.[42] Like Cuba, countries outside Europe and the former Soviet Union also faced major challenges. Many Third World governments traded with the Soviet Union, as it rarely required hard currency. Between 1982 and 1985, for instance, crude oil and middle-distillates, such as kerosene and diesel, represented 75 percent of India's imports from the Soviet Union, which was a major trading partner.[43] The collapse of North Korea's agricultural system in the 1990s can also in large part be ascribed to the loss of Soviet petrochemical imports. And yet, in global comparison, no country equaled Cuba in terms of its dependence on Soviet oil supplies.

SPARE PARTS

In addition to the oil cisterns gaping empty all over Cuba, a lack of spare parts contributed to the blackouts. This, too, was a question of geopolitics. By the late 1980s, the electricity and sugar industries were two of the most highly mechanized sectors in the Cuban economy, and most machinery originated from the Soviet Union, Czechoslovakia, and East Germany. When CMEA disintegrated in 1991, Cuban engineers quickly started to suffer from a shortfall in replacement parts, facing interrupted supply lines and deficient stockpiles. Fidel Castro pointed out that the hard currency the island could still raise from sugar exports had to cover everything from spare parts to food, medicines, and raw materials.[44] As a result, the thermoelectric power plants and the grid infrastructure of the SEN deteriorated quickly. In 1993, Minister of Heavy Industry Marcos Portal León notified the National Assembly that the thermoelectric plants had been left without resources for maintenance over the past few years and that this would remain the case throughout 1993.[45] The CTC's newspaper *Trabajadores* wrote that maintenance and repair was left to bricolage—to the electrical workers' "technically surprising maneuvers."[46] The lack of spare parts both increased the grid's susceptibility to blackouts and increased heat losses in transmission when the electric current met higher resistance in worn transformers and metal wires, thus increasing the SEN's systemic oil dependence.

A further complicating factor was the temporal pattern of electricity consumption across the grid: everyday activities in homes caused the system to blackout. "While rhythms of electricity use can be seen in particular technologies and activities," as Gordon Walker writes in his work on energy and rhythm, "for large-scale electricity systems the millions of instances of switching on and off that occur in specific devices coalesce (in quite an extraordinary way) into the aggregate rhythm of the moment-by-moment load on the system."[47] Industrialization and intensified energy consumption had been seen as preconditions for historical socialist progress, but the residential sector still accounted for 30–40 percent of Cuba's total electricity use in 1990. The proportion increased further when industrial and commercial activity declined during the special period.[48] This sectoral pattern meant that the demand for electricity—or the load on the system, to use engineering terminology—gave rise to a sharp peak at dusk, in the way Odalys in UNE alluded to. At home after work, people switched on the lights and their fans, opened their refrigerators, and tuned in the evening news. On average in 1999, the load on the SEN increased from about 1,550 MW at 4 p.m. to 1,850 MW at 7 p.m.[49] Quotidian activities in individual households had dire consequences on the grid scale.

For economic reasons, engineers try to maintain a high load factor. This implies that the average and maximum loads on the system are approximately equal over a given period. With a low load factor, it is necessary to keep generators on standby to be switched on during peak hours. Like all machines, these require maintenance and fuel, and usually, they are also the least efficient generators, consuming extra energy. During the special period, however, the shortage of fuel meant that the maximum load during peak hours exceeded the available standby capacity. Thus, around 5 p.m. in the average 1999 afternoon, the distribution circuits would blackout as the load rose above 1,700 MW, exceeding capacity.[50] Many weekly blackout timetables published in *Guerrillero* indicate that two or more circuits would be shut off by the utility in the evening when the demand would be at its highest. In this way, it tried to minimize the risk of a larger blackout, orchestrating the interrelated rhythms of supply and demand, even though the load on the SEN still often exceeded available supply with systemically detrimental effects.[51]

To reduce the load on the system, the government emphasized the importance of a trope well known since the Energy Forum in 1984: that of *ahorro*, or energy saving. Next to supply rationing, it targeted the demand side of the energy system. Reinier, whom I stayed with in Pinar del Río, recollected that one of the first tangible consequences of the special period was that the government emptied the shops of electrical appliances. They took away "the hotplates, the air conditioners, everything that if they sold them, they would increase the energy consumption," he said. Thus, it was impossible to get hold of new appliances through the formal distribution systems, whether the state shops, the rationing system, or the CTC's *emulación socialista*, rewarding vanguard workers with material goods.[52] In 1993, when the government legalized the US dollar as a means of payment, it also set up dollar stores in which a range of imported electrical appliances could be purchased. However, since the United States banned family remittances from being sent to the island, few Cubans had access to hard currency at this time. This left the majority of the population without access to new consumer goods.

Fidel Castro legitimized the withdrawal of appliances from the state shops in his speech to the Fourth Congress of the PCC, again rehearsing a concept of energy justice. If the socialist state was unable to supply energy, it would be irrational to distribute appliances that increased the load on the SEN, he argued: "Already since the end of 1990 we have had to limit the sales of televisions, radios, refrigerators, because if we should have to ration electricity, it doesn't make sense to continue distributing electrical appliances."[53] "So, *imagínate* [imagine for yourself]," Reinier said to me years later, "there were no appliances to get hold of except the ones you already had in your house. And so, you had to invent things and repair things."

CRISIS IN THE MASTER NARRATIVE

The SEN's physical extension and its institutional apparatus both coordinated and justified the socialist state's presence in space. When Cubans interacted with this centralized energy infrastructure, using appliances that converted electricity into the energy forms they desired, they were also hailed as socialist citizens. With energy use, José and Beatriz and Reinier and Anabel entered into the relational configuration of the Revolution's infrastructural form, both as energy users and as members of the socialist state. In the revolutionary narrative, the moment of human-infrastructure interaction signified revolutionary progress and increased political efficacy. In the early 1990s, however, the blackouts and the lack of cooking fuels undermined the narrative embedded in the energy system. The material realities in people's living rooms grated with the Revolution's historical vision of socialist modernity. Indeed, the "de-grown" environment had far-reaching cultural implications.

To normalize the friction between everyday experiences and the revolutionary narrative in both leadership and household discourses, the notion of the "special period" became a key discursive device. The special period had been a concept in Cuban politics well before the collapse of the Soviet Union, emanating from national defense planning in the 1980s. It denoted a situation in which Cuba would be directly or indirectly invaded by the United States, forcing the country into a state of emergency and national siege. When signs appeared of imminent political change in Eastern Europe and the Soviet Union, Fidel Castro explained the implications of this "special period in times of war" to the Fifth Congress of the Federation of Cuban Women (FMC): "The special period has been studied, analyzed, and prepared for in case of a total blockade of the country. . . . [We have defined] which measures to take in those conditions where absolutely nothing could come from abroad [and] what to do if military measures and actions are added to these measures; we have considered the plans,

including what to do in case of an invasion and occupation of the country."
With the socialist bloc's disintegration on the horizon, however, the strategy was redefined: "It would bring us a *special period in times of peace* if
very serious problems arise in the USSR and we cannot receive the supplies that we are getting from the USSR; among other things, the energy
supplies, which are so important to a country where the level of life and
development is based on a consumption of 12 million tons of oil."[54] Later
in 1990, Fidel declared in his speech to the CDR that "without a doubt, we
are now entering that special period in times of peace."[55]

From this point onward, the "special period" became the all-dominating
metaphor for life as it unfolded in the 1990s in both official and everyday
discourses. As a metahistorical descriptor, the concept incorporated difficult everyday experiences in a larger narrative, becoming a category of
experience. Among Cubans and foreign commentators, the special period
has subsequently been represented as a time of waiting; as a "perpetual
meanwhile" and an "irresolute transition."[56] And yet, to understand the
notion of the special period within the narrative logic of the Cuban Revolution, Fidel Castro's rendering of the crisis as a "special period" must
also be interpreted in relation to the dominant concept of history under-
pinning the revolutionary narrative.

In numerous public events, Fidel Castro described the special period as
a transitory phase in Cuba's longer revolutionary trajectory. This trajectory
was marked by the event, or triumph, of the Revolution in 1959, which liberated Cuba from the yoke of US imperialism. The Revolution brought
Cuba from underdevelopment in the neocolonial era into an epoch holding the potential for national development. In this new era, the possibility
of equal exchange with the Soviet Union enabled development on historically fair conditions and legitimized the return to sugar monoculture. Facing an ever more strained relationship with the Soviet Union in the late
1980s, Fidel Castro argued that the three decades of socialist trade thus
had been part of a period in Cuba's revolutionary history when development was possible. The end of equal exchange, however, now undermined
the Cuban Revolution as a historical epoch of socialist development: "This
historical epoch has not ended," he argued in 1990, "only a part of this in-
evitable historical period has passed."[57] In the post–Cold War geopolitical
order, Cuba was entering a temporary "special period" in which blackouts,

a lack of food, and economic austerity were difficult but historically necessary prices to pay for the Revolution to survive. Fidel could therefore tell
the FMC Congress, "If we remain five years without building one house,
well, if this is the price to save the Revolution, [then] we remain five years
without doing it, without building one daycare center."[58] As a category of
experience, the special period made difficult everyday experiences part of
a historical interlude in the longer temporality of revolution.

The intellectual historian Reinhart Koselleck argues that the modern
conception of time rests on two pillars: first, a formation of "collective sin-
gulars" and, second, a temporalization of these. In the modern era, time
and space have merged into a universal human history in which different
peoples are seen to be at different stages of the same historical path—the
path toward a modernity defined by a privileged Euro-American experience. All possible individual histories have merged into a universal collective singular History. In the meantime, the notion of the Revolution,
as a complementary collective singular, has introduced a coefficient of
movement in History. When a revolutionary event thrusts humanity into
a new epoch, obliterating the archaic social order of old, the Revolution
establishes irreversible temporal order.[59] In these terms, the SEN can be
seen to have produced a temporalized techno-material space holding the
promise of social change in the postrevolutionary period. However, the
breakdown of infrastructure in the special period undermined the integ-
rity of this space, fracturing the infrastructural form of the socialist state.
Historically, then, the Soviet Union's collapse and the United States' reinforced blockade together constituted a historical reaction to Cuba's in-
evitable march forward—a counterrevolutionary obstacle to be overcome
through national struggle.

The government's attitude toward the nuclear program demonstrates in
a particularly clear way how the special period was conceptualized as an ultimately transient historical phase. On September 2, 1992, Fidel Castro announced that the program would be indefinitely suspended. At this point,
approximately 6,500 people worked at the construction site in Juraguá, the
civil engineering aspect of the first reactor had been completed to 90 percent, and the investments exceeded 1.1 billion dollars.[60] First addressing the
workers in Juraguá and then a mass rally in Cienfuegos, Fidel explained
that Russia was demanding a renegotiation of the bilateral agreement for

the project. The one-year plan signed with Soviet officials in December 1990 had included supplies for the plant, but the Russian government only offered credits for a limited range of items, and Cuba would have to import essential components from a third country without Russian guarantees. Rubbing salt into the wound, the Russians also insisted on Cuba paying for shipping and that trade should be based on world-market prices.[61] Unequal exchange therefore made Cuba unable to meet the new contractual terms. Crucially, however, the construction of the nuclear plant was not canceled but suspended. A small workforce was left on site for basic maintenance. The suspension of the project was a great tragedy, but construction would resume as soon as the country returned to the development path, following the extraordinary special period.[62]

"RESISTANCE" AND "INVENTIVENESS" IN URBAN HOUSEHOLDS

For the Cuban people to reembark on its revolutionary path, the government called on the population to live up to a number of revolutionary values. These were part of a scale-based narrative, defining the moral code of a national, revolutionary subject. Addressing the FMC Congress, Fidel Castro explained the situation in these terms:

> War or not, special period or not, this is the most important moment in the history of our country and one of the most important in [the history] of the world; the moment in which it is decided whether all revolutionary flags are folded and a giant counterrevolutionary wave seizes the world for a long period of time, or whether we *struggle*, we *resist*, and we set an example and do what needs to be done. And we can keep up those flags, we can defend them, under whatever circumstance, with war or without war, with special period or without special period.[63]

The calls to "struggle" and "resist" in part resonated with the war rhetoric associated with the special period. As a concept of national defense, it implicitly identified an external enemy as the source of hardship—a counterrevolutionary adversary that could be fought. The call to arms, in turn, resounded in the history of Cuban anti-colonial struggle. In the revolutionary narrative, the M-26-7 represented the most recent and ultimately victorious expression of the struggle for national independence. The special period, however, required a new battle for the people to reclaim its sovereignty and to secure the Revolution against the Reaction.

Still in 2015, words like *resistir* (to resist), *inventar* (to invent), *luchar* (to struggle), and *resolver* (to resolve) echoed through memories of the special period. Rosa, Reinier, and I had just cleaned the house after an unexpected summer storm had rolled down Calle Martí, Pinar del Río's high street. A barrio north of Martí was still without electricity after a tree had fallen over the lines, and pickup trucks displaying the UNE logo sped down the road. "It shouldn't really be called the *período especial* [the special period], but the *período infernal* [the infernal period]," Reinier half-joked as we caught our breath. "'Special' sounds like it is something good. But really, it was an inferno." "These days we sit around and laugh at these memories," Rosa said. "But you know, really, how did we survive?" She then paused to consider her own question before continuing: "But we did survive. We *invented* and we *resisted.*" As a category of experience, the special period systematically attained meaning in relation to these verbs (Figure 7). Many ethnographic studies also demonstrate how they invoke critical abilities in contemporary Cuban everyday life.[64] Now articulated on an individual or a family-based scale, they are ways of speaking about activities and transactions that either solve immediate practical problems or through which a person generates extra income to make a living.

The ethnographic literature has mainly focused on the latter aspect, on strategies for a person to earn money or acquire desirable things by calling in favors, selling lollipops or almonds on the street, or by turning to street hustling or prostitution (*jineterismo*). Kathy Powell argues that survival in the special period depended on informal networks of solidarity with relatives and friends, abroad and on the island, as well as with local representatives of the state. Nevertheless, the scarcity of resources limited people's ability to keep up reciprocal social relations, which gave rise to intricate spheres of exchange articulated in terms of "inventing" and "resolving." David Forrest writes that the need for informal channels to access resources—and not least hard currency after the legalization of the US dollar—predominantly benefited the white population in their efforts to "resolve" and "resist" because of their closer relations with people

Figure 7. "Only those win who struggle and resist." On the highway between Havana and Pinar del Río, a bridge mural reproduced the official *lucha* narrative (*"Solo vencen los que luchan y resisten"*). The person underneath the bridge was hitchhiking—a mode of transport institutionalized during the special period to make use of all available fuel. In theory, all state-owned cars had to pick up hitchhikers. Waiting long hours for a car to stop could be described as a form of everyday "struggle." Photo by Gustav Cederlöf.

who gatekept resources. Instead of *socialismo* (socialism), there was now a system of *sociolismo* ("buddy socialism") quickly causing racialized inequalities to grow.[65]

However, words like *invent, resist, struggle,* and *resolve* also had a more literal and often overlooked meaning. On one occasion, Reinier brought me his daughter's mp3-player. He wanted me to listen to a song by the Havana-based *nueva trova* musician Tony Ávila, which he thought brilliantly captured the essence of the special period. "Regalao murió en el 80" (Regalao died in 1980) begins with witty references to events during the Soviet era: the unsuccessful Ten Million Ton Harvest, the Mariel boatlift, and the abundant supplies on the *libreta*. But then the crisis ensues, which leads up to the song's title—the death of Regalao. Regalao is a personification of the verb *regalar* (to give or gift), and much like Powell argues, the scarcities of the 1990s put a strain on established spheres of exchange and the meaning of generosity. But Ávila's text also characterizes the technical innovations of the time. Cubans' inventions were ingenious but far from cutting-edge solutions, he writes, as "we gave it pumpkins" (*le damos cala-bazas*). Reinier explained the idiom *dar calabazas* as signifying the ability to resolve things with elementary, mundane solutions. High-tech scientific advances were not applicable in Cuba's special period. "*We gave it pumpkins* means that we had to find quick solutions; find traditional, elementary solutions; ingenious solutions that no one would think of," he said. "You had to resolve things by adapting to the new times, without having access to scientific and technological advances." The special period called for bricolage out of the ordinary, which translated into a practice of "resisting" the crisis and "struggling" against reactionary forces.

In numerous conversations, one anecdote was repeatedly retold to ex-emplify, or perhaps epitomize, what *inventar* and *resistir* meant in the special period. The story would usually start with the person's family that, like most families, never could get hold of a new electric fan. Fortunately, they had an old Soviet washing machine—an Aurika. The machine had two parts, one for washing and one for tumble drying, and the tumble dryer was a poor build that frequently broke. This meant that, at no loss, the machine could be cut in half, and if you put rotor blades on the tumble-dryer motor, you were left with a fully functioning washing machine and an electric fan. "The motor had immense power, and you had to put large blades on it," one person recalled. "It was like an airplane!" Vilma, who worked as a librarian, showed me a short film by the Cuban filmmaker Luis Lago in which a man is haunted by his homemade fan. He repeatedly goes to check on it, but the fan ultimately kills him in his sleep as it vibrates across the floor. The case of the monstrous fan is also the example that Tony Ávila chooses to exemplify Cuban inventiveness. You cannot find a problem without a solution, he sings; if you need a fan, just split your Russian washing machine in two.[66]

Importantly, the ability to invent or repurpose things was not a universal skill; rather, it was a characteristic of the Cuban national ethos. Ávila suggests that not even the Japanese or the Americans in NASA would be able to come up with the Cubans' innovations. One day, I asked Sergey how his mother had cooked with charcoal when all they had was a kerosene burner. "Ah!" he replied. He then stopped talking about his mother and began talking about "the Cubans" as a third-person collective singular: "The Cubans are tough; they *invent* what they need. The Cubans invented stoves. For example, you could take a tin—you know they come in these small sizes, but also larger—and you'd take a big one and then make two holes on the sides toward the bottom to get a draught, and then fill it up with charcoal or wood or whatever. And then you put a grate or something on top, and there you go. See—the Cubans invent." The friction between everyday experience and the revolutionary master narrative was negotiated in the act of low-tech innovation, but it was also articulated satirically. In a cartoon Anabel had kept, a man calls his wife over to the TV. A coat hanger is fitted as an antenna and Fidel Castro appears on screen in a chef's apron: "Hurry, darling," the man shouts, "Fidel is going to explain how you make *mamey* milkshake without *mamey*, without milk, and without a blender!"

Thus, memories of bricolage—ad hoc innovations that enabled long-established practices of energy use to take place in the midst of black-outs and decrepit appliances—were articulated as instances of Cubans' remarkable ability to invent and resolve things. Marisa Wilson argues that the verbs *inventar* and *resolver*, in the way they coarticulate *resistir* and *luchar*, allowed people to reconcile everyday experiences with the official revolutionary narrative: "The ability to 'resist' is a revolutionary value that allows people to justify everyday difficulties."[67] In terms of energy use, the *lucha* narrative therefore worked on two linked scales so that lived experiences of consumption resonated with a national ethos, and national "resistance" was acted out in everyday life. An intertextual play was at work when the activities described as inventiveness and resistance tapped into the revolutionary narrative of anti-colonial struggle and the war-rhetoric associated with the special period.

In sum, the material transformations precipitated by the Soviet collapse, such as the reduced availability of oil, the shortages of spare parts, and the alarming load factor in the SEN, compromised the infrastructural form of the socialist state not only in political economic but also cultural terms. However, the notion of the special period and the *lucha* narrative worked to normalize the crisis. To "invent" and "resist" the special period did not signify resistance *to* the Revolution but rather resistance *within* it in order for it to survive.[68] While the economy de-grew materially, this smaller metabolism was experienced by socialist citizens struggling in an interim special period. "*Luchar* can imply struggling within the Revolution," Reinier clarified, "because the state has to struggle against the blockade and imperialism." Interestingly, the ability of "the Cubans" to invent, resist, struggle, and resolve was the ability of a nongendered, non-racialized subject. And yet, since they did not share the same spaces for human-infrastructure interaction, men, women, and Cubans of different color experienced the crisis differently. These social inequalities were disarticulated in the intertextual play between discourses working at the household and national scales. The special period required Cubans to live up to the values that characterized them as a singular revolutionary people. The wider narrative, making practices of energy use meaningful, was one of collective struggle in the face of adversity.

4 Socialist Redistribution and Autonomous Infrastructure

The US oil blockade in 1960 and the problems in the Soviet oil industry in the 1980s had kept Cuba's oil import dependence in the public eye. But in the 1990s, the collapse of the Soviet Union created an immediate need for collective action in the face of acute oil shortages. A critical juncture like the Soviet collapse, Kent Calder writes in his study of Eurasian energy geopolitics, is "a historical decision point at which there are distinctive alternative paths to the future."[1] The national energy crisis rose to the top of the agenda at all levels of the Cuban political and economic system for debate. In 1992, the state administration began developing a new cohesive energy policy. Raúl Castro, then minister of defense, headed the process that engaged several ministries, the National Energy Commission, the central planning board JUCEPLAN, and the Cuban Academy of Science. The Council of Ministers then approved the Program for the Development of National Energy Sources (hereafter, the Programa Energético) in May 1993 before the process formally concluded in late June that year when the National Assembly ratified the program.

The Cuban newspapers show how a larger exercise in the Poder Popular, the executive and legislative branch of government, dovetailed the development of the Programa Energético. In 1993, the PCC and the CTC announced that meetings were taking place in the Popular Councils—the sub-municipal tier of government—as well as in factories, farms, and universities. The population was canvassing ideas for reduced energy consumption.[2] In Pinar del Río, the Provincial Assembly requested individuals and teams to submit prototypes, photos, and posters with designs that would contribute to the conservation of energy. The submissions would be displayed in the assembly hall, and the mass organizations promised "special awards" to the best contributions.[3] Among a host of other events, engineers in Pinar del Río shared ideas in a workshop on how to incorporate energy resources available in the region in the local energy systems.[4]

As part of the political process, the municipal assemblies first convened, discussing local experiences, followed by the provincial assemblies and then the National Assembly. As experiences moved up the institutional hierarchy, the deputies stressed that solutions had to be found locally but that the political process would help to "generalize the experiences," enabling them to be transferred "to other sites."[5] At the end of the exercise, the process—at this point spoken of as a completed entity—gave evidence to the thoroughness of the Programa Energético's appraisal. Jaime Crombet, the vice president of the National Assembly, testified to parliament that "the inclusion of the energy theme in the legislature . . . [has] counted on the participation of the masses."[6] The political exercise thus legitimized the Programa Energético as a product of great popular involvement. "All the information that we have at our disposal from various municipalities and provinces," the PCC's newspaper *Granma* concluded, "gives faith in the profundity with which the issue has been considered."[7]

The state administration drafted the Programa Energético around two central propositions: first, that all imported fuels should be substituted with nationally produced energy resources and, second, that the efficiency of energy use should improve in the state economy. Many energy sources together, even if available only in small quantities, would reduce the need for imported oil, and higher efficiency would reduce the need for primary energy input in absolute terms. The day before the National Assembly, the newspaper *Juventud Rebelde* explicated the two propositions to its readers: "Cuba is not an oil country; neither do we have a great hydropower potential, nor coal, nor large forests, nor electronuclear [power plants]; nonetheless, we can count on certain quantities, modest or significant, of some natural elements: oil, wood, peat, biomasses, some rivers and streams, winds, a lot of sun and sugar cane. An adequate combination of these elements, combined with the efficient exploitation of the processes of combustion, transmission, distribution and energy use must take us on to the right path."[8]

Moving forward, the Programa Energético became the guiding framework for attempts to resuscitate the ailing state economy, seeking to reestablish the political economy of centralized socialist distribution. However, in focusing on local solutions, the Programa Energético also opened a space for the population to set up new infrastructural systems, relying on energy sources available in proximity to them, such as charcoal, wind, and biogas. While the new policy was a strategy for socialist development, it is easy to attribute these alternative energy systems to a smaller scale degrowth ideal, allowing human activities to continue with significantly reduced energy and carbon intensity. Indeed, the proliferation of nonstate-based infrastructural systems transformed established patterns of distribution and connectivity, and in doing so, they contradicted key political priorities in the socialist state. Rather than revive the state's infrastructural form, the alternative solutions enabled individuals and cooperatives to withdraw from it as they provided the conditions of possibility for increased local autonomy. In the special period, infrastructure again became a medium, and energy use a moment, at which the future political direction of the socialist project was negotiated.

RESUSCITATING THE INFRASTRUCTURAL STATE:
FOREIGN CAPITAL IN THE SEN

Confronted with the Programa Energético, managers in the state economy started work to restore the levels of energy throughput in the centralized infrastructures. UNE's primary objective was to diversify the energy mix in the SEN and in that way curtail the nationwide blackouts. This was not only an issue of replacing Soviet oil with alternative energy sources but also of territorializing a new regime of energy supply on the island. National electrification had long increased Cuba's dependence on fossil fuels, and drawn from underground, these energy sources contributed to the production of the particular spatial patterns of economic and social activity that integrated the socialist state overground.[9] The political and environmental challenge now facing Cubans was to replace oil imports either with fossil energy extracted from the national subsoil or renewable energy harnessed from contemporary eco-productive space. Thus, the SEN had to go through a scalar reconfiguration so that instead of relying on transatlantic trade, energy resources were extracted from national territory.

As noted in chapter 1, Cuba's proximity to the Gulf of Mexico had aroused speculations about the existence of vast oil deposits within Cuban borders since the early twentieth century. Even so, little prospecting ever took place after the Revolution when oil arrived aplenty from overseas. In 1990, Cuba's petrochemical production amounted to 0.67 Mt, output consisting mainly of naphtha. But the Programa Energético radically changed the government's approach to the industry: the Ministry of Heavy Industry (MINBAS) and the Ministry of Foreign Trade (MINCEX) started inviting international oil companies to prospect for oil in the Cuban offshore, working jointly with Cuba's national oil company, Cupet, in doing so.[10] By 1995, Sherritt International, Pebercan, British-Borneo, and several other companies had helped Cuban production reach 1.47 Mt, which peaked at 3.68 Mt in 2003.[11]

The supplies of national crude were used directly, without refining, in the SEN's power plants. Refining is itself a highly energy-intensive process, and by bypassing the process, MINBAS eased the pressure on the national energy budget. Nevertheless, compared to Soviet fuel oil, the Cuban crude was both "sour" and highly viscous, having a high sulfur content and an asphalt-like texture. Designed as they were in the Soviet Union and Czechoslovakia, the thermoelectric plants were poorly adapted to the heavy crude, and it aggressively corroded the steam boilers and left heavy residues behind. As a result, the power plants required more frequent maintenance, which led to longer blackouts, both scheduled and unscheduled. In January 1993, UNE carried out enameling works to make the steam boilers more resistant to corrosion, at which point the newspaper *Trabajadores* reported, "The forces are brought into tension, because in the same instant that the use of national crude is increased, the maintenance works become more frequent and the number of parts to replace and repair increases."[12]

The main obstacle to a major technical overhaul of the SEN was the lack of finance. Short on hard currency, the state required direct investments or aid from abroad to undertake any retrofitting, and it needed partners who were prepared to bypass the US economic blockade. While it is difficult to access primary sources detailing government deliberations from the time, MINCEX secured a short-term credit from the French insurance company Coface in 1997, allowing it to adapt the "Antonio Maceo" plant in Renté to the use of Cuban crude.[13] This was the first in a series of projects with the same purpose, and by the turn of the millennium, UNE was generating between 70 and 90 percent of all electricity from national crude in plants customized to Cuban fuel. The electricity supply remained highly unstable, but this intervention effectively brought an end to the scheduled blackouts having plagued the population for over a decade.[14]

To further restore the SEN, the government made the electricity sector part of a more wide-ranging structural transformation of the state economy. In the early 1980s, the Council of State had authorized foreign investment on the island via joint ventures by which the state owned at least 51 percent of a company while a foreign investor was offered the remaining shares in return for a capital contribution. When Cuba's relations with the Soviet Union deteriorated, the government actively sought capital by these means in order to develop the tourism industry as a potential source of hard currency income. The Spanish hotel chain Sol Meliá became the first major foreign actor in the Cuban economy, and in the late 1990s, two joint ventures were set up with the Spanish utility Aguas de Barcelona to manage the water supply in Varadero and Havana. In a detailed study, Karen Cocq and David McDonald demonstrate how the water enterprise in part resulted from Sol Meliá's wish to secure a stable water supply to its hotels rather than more immediate Cuban domestic priorities.[15]

In 2003, a full decade after the Programa Energético was adopted, a first joint venture was established in the electricity sector. UNE, Cupet, and the Canadian mining company Sherritt International formed the new utility Energas, each holding one-third of the shares. Sherritt had been active in Cuba since the early 1990s when it agreed to purchase Cuban nickel ore, which previously had been exported to the Soviet Union. Sherritt partnered with the state mining company Cubaniquel in 1994, and Cupet then contracted Sherritt for oil production in the Gulf of Mexico.[16]

The formation of Energas was predicated on the possibility of utilizing natural gas extracted from Sherritt's oil wells, and Energas built three gas-fired power plants next to Sherritt's oil concessions to generate electricity.[17] Energas was majority owned by the socialist state, but just as natural gas technology had been integral to destabilizing the paradigm based around thermoelectric generation and vertically integrated utilities in the United States, Energas broke UNE's monopoly in the electricity supply sector.[18] While Sherritt expanded its venture in Cuba, the Cuban government gained access to foreign capital, and Energas—acting as an independent enterprise—displaced demand from UNE's oil-fired thermoelectric plants in the SEN.

SUGAR AND ELECTRICITY

Alongside domestic oil and gas production, the Programa Energético projected far-reaching changes in the sugar industry. Prior to the Revolution, many *centrales* had been important generating nodes, supplying electricity to mills, railroads, and nearby towns. With the construction of the SEN, however, generation moved from geographically dispersed locations to centralized thermoelectric plants where scale effects provided economically more efficient generation. Even so, sugar retained an important role in the national energy system as it was traded for oil. In the post-Soviet geopolitical landscape, the government reasoned that products from sugar production also could be used directly in the energy system. Bagasse, cane leaf, and filter cake could be incorporated as use values in the system, diminishing the need for sugar as an exchange value on the international markets.

The Programa Energético specified that the sugar industry should "become self-sufficient and contribute to the grid" as soon as possible.[19] Bagasse was the most important fuel toward this end, as it was both energy potent and existed in abundance during the milling season. In June 1993, the Ministry of Sugar (MINAZ) planned for forty-two turbogenerators to be installed in its *centrales* to make use of bagasse and other cane biomass. Together with the generators already operating in some of the mills, the turbogenerators would feed electricity into the SEN.[20] Instantly, however,

MINAZ ran into problems. The sugar mills' steam turbines were all designed to work at low pressures, because when bagasse was regarded as a waste product, low turbine pressures allowed MINAZ to get rid of more bagasse in a shorter period of time.[21] While this served a practical purpose, it was an economically inefficient way of converting biomass into mechanical energy for the purpose of electricity generation.

Two more challenges impeded the transformation of the sugar industry into a net electricity producer. First, the industry itself was a highly oil-dependent production system. After mechanization—most notably during the 1976–80 five-year plan—it, too, relied on the transatlantic oil-for-sugar exchange mechanism. In Pinar del Río, with large *centrales* in Bahía Honda and San Cristóbal, MINAZ reported that it was unable to meet its production quota for 1993.[22] In the country at large, the *zafra* had reached 7 Mt in 1992 but only amounted to 4.2 Mt in 1993. The lack of agrichemicals, requiring great quantities of natural gas and other fossil fuels in production, left the cane fields unfertile, even as MINAZ lacked fuels for transportation. Fidel Castro reported that almost 1 Mt of cane was left cut but unmilled in the fields at the end of the *zafra*—cane worth approximately two hundred million dollars. While fuel had been assigned to the mills, it often failed to reach them due to transportation problems higher up the supply chain.[23] Thus, a smaller harvest, resulting from a lack of petrochemicals, led to less bagasse, which in turn led to a lower electricity output.

Second, the sugar sector proved to be highly energy inefficient. Efficiency had been an issue of little concern when the sugar price was pegged to the price of oil, and sugar output gave instant access to oil. When Cuba instead traded wholly on the international markets, sugar generated diminishing returns.[24] According to one study, the industrial machinery was old, inadequately maintained, and at times ill designed. This created a situation in which it was cheaper for MINAZ to import sugar than to produce it on the island.[25] As a result, the government decided to close 71 of the country's 156 sugar mills in 2002—a choice effectively putting an end to 150 years of sugar monoculture in Cuba. The remaining *centrales* were all synchronized with the SEN but remained net importers of electricity. Consequently, the sugar industry was unable to provide a new source of low-entropy energy to the broader state infrastructures. Despite parallel investments in smaller hydroelectric power plants, wind turbines, and photovoltaic solar panels—the latter mainly contributing to off-grid electrification in rural areas—the proportion of renewables remained small in the SEN. In 2014, 95 percent of all electricity was still generated from fossil fuels.[26]

COMBUSTION AND THE POLITICAL ECOLOGY
OF RENEWABLE ENERGY PRODUCTION

Next to electricity, many everyday practices of energy use relied on fossil fuel combustion to directly utilize the heat it generated. At home and in workplace canteens, Cubans were used to dealing with kerosene and LPG for cooking; in industry and transport, they relied on fuel oil, diesel, gasoline, and naphtha. As the state's reserves of cooking fuels ran out, these were often replaced with locally produced wood and charcoal. Magdalena, a woman around sixty in Pinar del Río city, recalled how they would carve wood into sawdust and burn it. Rosa explained how "people even cooked with wood in the high-rise buildings [lit. the twelve-story buildings]—*¡imagínate!*—they made stoves that they kept on their balconies. And many houses didn't have window blinds. People burned the wood and didn't care if the windows were left open, because that way the smoke could get out."

In the province of Pinar del Río, the politburo advocated the installation of wood burners in the canteens of state enterprises and state farms. To guarantee a stable supply of wood, energy forests would be planted "behind or in front or on one side or on the other of every burner, in every canteen." It was up to the Communist Party and the union cells to "demand that it is achieved."[27] A few weeks later, *Guerrillero* reported that the Municipal Assembly in Candelaria had decided to plant energy forests next to all work centers using wood or charcoal to make them self-sustaining.[28] However, felling to secure wood supply was not uncontroversial. A delegate from the municipality of La Palma argued in the National Assembly that wood cutting had to be regulated to protect the forests. The endemic Caribbean and tropical pines as well as Cuban oak dominated the forests worthy of protection in Pinar del Río. The delegate stressed that wood and charcoal outtake was advisable only "in places close to where they will be used."[29]

There were also initiatives to substitute rice straw and peat for oil. The Ministry of Agriculture (MINAG) in Pinar del Río reported that eight rice-drying ovens would run on straw by July 1993.[30] A study brought before the Provincial Assembly indicated that almost four thousand tons of oil could be "saved" in the province from the burning of rice hull.[31] A feature in *Granma* related the successes of rice-straw burning in the province of Sancti Spíritus where straw from a normal harvest had replaced 1,500–1,600 tons of oil in ovens for cereal drying. This was only a mere 20 percent of the total rice by-product, and straw was also sent for combustion in a nearby bakery, a launderette, a rice-precooking industry, and a cement factory.[32] In addition, the first secretary of the PCC in Pinar del Río called for accelerated peat cutting in the bogs around San Luis and Sandino, "with a view to offer the population another fuel."[33] The newspapers noted that peat was the most abundant fossil fuel of all in the country. The deposits in the Zapata Swamp alone accounted for two hundred million tons-of-oil-equivalents.[34]

Through measures such as these, the government attempted to replace previously imported oil on a joule-for-joule basis. Reflecting the Programa Energético's vision for a nationally territorialized energy system, it used domestic oil, natural gas, bagasse, wood, charcoal, rice straw, and peat to fuel the energy systems. Everyday energy use could then proceed unaltered via the centralized distribution systems, whether this energy use constituted electricity production in a thermoelectric plant, sugar crystallization in a sugar mill, or cooking in a home kitchen. Through a diversification of energy sources within the energy system, the state infrastructures remained largely intact. While energy previously had been supplied in the form of imported oil, the infrastructural form of the state now incorporated energy in the shape of biomass harvested directly on Cuban soil and fuels extracted in the Cuban offshore.

Nevertheless, these alternative energy sources did not accumulate in a social vacuum. Like the supply of Soviet oil, the flow of energy was firmly situated in historically and geographically determined social relations in which the costs and benefits of production were unevenly distributed among social groups. In chapter 3, we saw how charcoal was used in households, and charcoal provides a telling example of how land-based energy resources generated conflict. In a general sense, it is difficult to quantify the land intensity of charcoal production with precision—it varies with the wood species, soil fertility, elevation, climate, the efficiency of the kiln and other things. Even so, Vaclav Smil estimates that if global steel production were fueled with industrially produced charcoal instead of coke (or another fossil fuel), an area equivalent to half the Brazilian Amazon would be needed annually.[35] Faced with the high land intensity of charcoal production, it is questionable if a biofuel such as this can sustain a high-energy economy without generating land-based conflicts, regardless of whether it is produced in a socialist, capitalist, or other system.[36] The violence associated with rapid land-use change may be an argument for degrowth in itself.

At the smaller, household scale, the socialist state distributed charcoal as a use value to the population as far as stocks permitted. Next to calls for energy-forest plantations and peat cutting, Party officials also asked for intensified charcoal production in Pinar del Río. To showcase the state's efforts, *Guerrillero* published a portrait of a vanguard charcoal producer named Mongo Dopio in 1993. Mongo Dopio had once kept six kilns alight on his own. He had also committed to producing three thousand sacks of charcoal in one year for which he would not use a single drop of oil and rely entirely on animal traction. Mongo Dopio explained how the state recognized the importance of his work and incentivized production by giving "us coalmen an additional quota of food, cigarettes and cigars." He—and with him the Communist Party's newspaper—then confirmed the sincerity of the endeavor: "Listen, journalist, don't forget to write that Mongo Dopio will fulfill his commitment of three thousand sacks that he has for this year, because we coalmen know the importance that our work has today more than ever" (Figure 8).[37]

Even as the state relayed its efforts to increase production, charcoal created political tensions. The conflict was linked to the horizontal extensiveness of charcoal production. With the second Agrarian Reform Law of 1963, the revolutionary government nationalized landholdings exceeding five *caballerías* (67 ha). Peasants holding smaller areas retained their land and were subsequently identified as members of the peasantry based on their title. However, as noted in the context of electrification, the peasantry was often seen as a social class of the prerevolutionary *antes*, a class set apart from the socialist project. In the army newspaper *Olivo*

Figure 8. Small farmer producing charcoal in a kiln. (The photo is unrelated to Mongo Dopio and the *Guerrillero* article about him.) Photo by iStock.com/Nikada.

Verde, Che Guevara argued that the peasants, who in contrast to the proletarian sugar-workers toiled to enrich themselves, had "a petty-bourgeois spirit."[38] Almost everyone who remained without a connection to the SEN (5.8 percent of the population in 1990) were also small farmers living in areas remote to the state.[39] Yet, in the early 1990s, rising urban demand for charcoal placed many small farmers in control of a highly valued energy resource.

When I sat down with Sergey at the local library, he explained that his mother had cooked with kerosene until the state's reserves ran out during the special period. She had then turned to charcoal, which she burned in a homemade stove made from a large aluminum tin. Sergey explained that almost all charcoal was sold on the informal market by small farmers. "The state held some reserves," he said, "which it distributed. But the majority came through people who sold on what we now call *cuenta propia,* although it wasn't called that back then." *Cuentapropistas,* or "people who work on their own account," are Cubans who run their own small businesses in the nonstate sector in the aftermath of Raúl Castro's economic reforms of 2011. "But this wasn't strictly legal?" I asked. "It was *por la izquierda* [lit. 'to the left'; on the black market]?" "Yes, it was *por la izquierda*, but the state didn't care, they allowed it."

Media reports from the municipal and provincial assemblies in Pinar del Río indicate that the informal charcoal market was a matter of concern to the government. In Candelaria, the Municipal Assembly decided to increase vigilance in state-run enterprises. Some individuals were producing charcoal from state-owned wood and then charged "abusive prices" for it, the assembly noted.[40] In the Provincial Assembly, the deputies also argued that the production of peat should be intensified to weaken the informal charcoal market: "In this way the underground commerce of charcoal, which has unpopular prices, is in part averted," *Guerrillero* reported.[41]

From a Marxist-Leninist perspective, the property rights of charcoal should be universalized in the socialist state, enabling centralized distribution. But in the early 1990s, the regulation of land use set out in the second Agrarian Reform Law created opportunities for peasants that defied the state's resource monopoly. Many peasants had forests growing on their land, sometimes as part of an agroforestry system. Others could swindle wood from state plantations. Many also had traditional knowledge of how to refine wood into charcoal in slow-burning kilns. Thus, formerly marginalized peasants controlled the primary resource and had command over the knowledge-intensive practice of refining it. While oil and bagasse were produced in spaces controlled by the state, the materiality of charcoal, coupled with the regulation of land use, enabled nonstate actors to produce this energy resource. The urban demand for charcoal, finally, provided them with the opportunity to trade it for a profit, capitalizing on charcoal as an exchange value. Since the Cuban peso was heavily inflated, this trade rarely involved money. Instead, charcoal was traded for items that urban residents had available, and as a result, many peasants were comparatively better off in material terms than the urban population.

Studies of Cuba's urban food supply systems have identified a similar dynamic. While small farmers lived in relative affluence from the food they produced on their own land, the loss of food imports from the socialist bloc hampered the state's ability to ration food in cities. The scarcity of state-owned food thus generated urban demand for peasant produce in both the formal and informal markets.[42] In terms of charcoal, Party officials argued that the informal trade was "unpopular" and "abusive," and possibly, it would even give rise to a petty bourgeoise that threatened the socialist state's hegemony over energy distribution. Charcoal, like other energy resources, was an energy source that the population should have access to on equal terms, not as a commodity but as an entitlement. While the government legalized free farmers' markets for food in the early 1990s, the charcoal market remained informal, even as the government allowed it to proliferate. The government was unable to meet the demand for cooking fuels through the centralized energy infrastructures, and it was unequipped to take on the peasantry at this time of heightened state vulnerability. The politics of land use inherent in biomass-based energy use thus empowered small farmers in the special period.

ENERGY CONSERVATION AND ECONOMIC ENERGY EFFICIENCY

The efforts to substitute national for imported energy resources went hand in hand with initiatives to reduce energy demand and the load on the state infrastructures. The Programa Energético identified two types of initiatives for energy conservation. First, *ahorro* again re-emerged as an ideological imperative. It had been a key feature of the energy policy adopted at the Energy Forum in 1984, and it legitimized the scheduling of blackouts and the withdrawal of electrical appliances from state shops. During the special period, state representatives repeatedly instilled the importance of energy saving in the population: the need for *ahorro* was rehearsed on billboards, in political speeches and newspapers. In Pinar del Río, to take one example, *Guerrillero* declared that "every man and woman in his or her workplace must become conscious . . . [of the objective to] produce more and with better quality and with less consumption of fuel." To "save energy" was "the unpostponable responsibility of every entity, household, [and] canteen."[43]

To reduce the demand for electricity, MINBAS launched a nationwide campaign, which it branded the Programa de Ahorro de Electricidad en Cuba (PAEC—Program of Electricity Saving in Cuba). The objective was to breed a new culture of energy use on the island—to "develop habits and customs of rational energy use" among the population.[44] In PAEC publications, MINBAS instructed the public not to keep air conditioners and refrigerators in direct sunlight, to paint walls in bright colors, to switch off the lights when leaving a room, and to use washing machines outside of peak hours.[45] As a continuation of PAEC, the Ministry of Education amended the school curricula to promote a "rational culture" of energy use in the younger generations, and MINBAS developed materials to be used in schools with the new curriculum emphasizing the importance of *ahorro* to the family economy, the environment, and the continued success of the Revolution.[46]

In a second kind of initiative, the government set engineers and scientists to work to increase the economic efficiency of energy use, making the conversion of one energy form into another more effective within a given energy system. The engineers communicated some of their ideas to the public in the national media. In *Granma*, scientists in Havana recommended that people using kerosene for cooking should place the liquid in an inverted plastic bottle and suspend it 60 cm above the fuel tube's entry into the cook stove. This would create a constant pressure on the liquid, which would improve the efficiency of combustion. If the bottle was installed in "only" 40 percent of the country's two million kerosene stoves, five thousand tons of kerosene could be "saved" in one year.[47] In another article, *Granma* instructed its readers that an average Cuban refrigerator could "save" 6–7 kWh per month—equaling 10–12 percent of its total energy consumption—if copper or aluminum wires, 1.2–2 mm in diameter and 8 cm long, were fastened on the radiator grille. Each wire should be wrapped once around the tubing and the ends twisted together. This "little bow" (*moñito*) of metal would expand the surface area of the refrigerator's condenser and thereby allow heat to dissipate more quickly. "This simple '*moñito*,'" the newspaper concluded, "is an economic novelty that gives discount to the global energy consumption, and in this direction every grain of sand counts."[48]

Similar solutions were found in industry. In the sugar sector, MINAZ attempted to increase the energy efficiency by taking better care of hot water and by reducing the loss of steam from boilers. Fidel Ramos Perera, the first secretary of the PCC in Pinar del Río, suggested that the steam generated in the *centrales* also could be used for non-sugar purposes before it condensed. He reported that the Central Agroindustrial (CAI) "Pablo de la Torriente Brau," a sugar mill east of Bahía Honda, had started to provide a sweets factory in the adjoining *batey* with steam.[49] Among the more spectacular attempts to make energy use more efficient were experiments to magnetize fuels in the transport sector. The idea behind magnetization was to realign the molecules in gasoline and diesel into an even pattern by letting the fuel pass through a magnetic field. This would make the hydrocarbon molecules bond more easily with oxygen during combustion, leading to a cleaner, more efficient combustion process.[50]

With energy saving measures such as these, the levels of energy consumption were reduced in both absolute and relative terms. Workers and managers experimented with new solutions to solve immediate problems. Government representatives promoted energy conservation, and scientists presented efficiency-enhancing innovations for the country to reduce the load on the energy infrastructures. At the same time, the government worked to replace the oil-for-sugar exchange mechanism with an energy regime territorialized on the Cuban islands, reconfiguring the political economic dimensions of the state's infrastructural form. A diverse range of nationally available energy resources was at the heart of these efforts, although the partial rescaling of the energy system did not rid it of international connections. Through joint ventures, the government integrated the socialist state with transnational capital while making sure to retain control over all enterprises acting in the national economy.

REINVENTING INFRASTRUCTURE: DIVERSIFICATION IN THE AGROINDUSTRY

All efforts examined up to this point were attempts to resuscitate Cuba's centralized energy infrastructures. However, alongside these efforts, many infrastructures were redesigned altogether. Workers and farmers did not try to find alternative energy sources within the existing systems or to make energy use economically more efficient, but rather, by entering into new infrastructural relations, they tried to make the centralized energy infrastructures redundant. In the agricultural sector, many farmers had long used electric motors to power their water pumps. Electricity allowed them to feed water across pastures and through irrigation systems, but in the early 1990s, many stockbreeders instead started using wind turbines to directly impel their water pumps. "The use of wind energy (windmills) has come back to life again," *Guerrillero* reported from Pinar del Río's Provincial Assembly in 1993, "and there are 264 water pumps operated by that system in the province, apart from another 150 centrally allocated [to be installed] this year."[51] The pumps required mechanical energy, and instead of converting electricity generated from fuel oil into this energy form, the wind turbines directly converted the motion of the wind into the desired energy form.

The Empresa Pecuaria Genética (EPG) "Camilo Cienfuegos" was a large dairy-farming enterprise located in Consolación del Sur, an hour east of Pinar del Río city. The company had a central administration area surrounded by farms incorporated in the larger organization. In total, there were thirty-two active windmills pumping water around the company's pastures. The EPG's energy manager explained, "To me, the windmills are the most important renewable energy sources, because they take water out to the cattle without using oil; they are constantly shooting out [*tirando*] water and are saving an immense quantity of fuel." After making a quick calculation, he and the workers listening in estimated that the company "saved" 72,000 L of oil per year with the wind turbines—"not a huge, but a significant quantity."

Government representatives in Pinar del Río also called for the use of gravity irrigation in the agricultural sector so that Earth's gravitational pull would replace electric motors to move water.[52] In the National Assembly, a delegate from Santiago de Cuba explained how water moved by gravity was used in his constituency to first spin mini-hydroelectric turbines and then for canal-based irrigation. There were also reports of farmers using mechanical energy from water turbines directly. For example, a delegate from the Escambray Mountains described how his constituents were using water turbines for coffee pulping, rice hulling, and animal-feed processing. It was also used in a staple factory.[53]

To dry coffee beans and cereals and to heat water, the use of solar heaters proliferated. A solar heater is a device that absorbs solar rays, concentrates the heat in a conductive medium such as a gas or a metal, and then transfers the heat on to a drying tray or to water passing through a

Figure 9. Solar heater on Carlos and Gladys's farm. Photo by Gustav Cederlöf.

pipe.[54] Like gravity irrigation and windmills, solar heaters had been in use in Cuba long before the special period, but in 1993, the state again started to manufacture the technology.[55] Carlos and Gladys were dairy farmers working for EPG "Camilo Cienfuegos," and they had a Chinese solar heater installed outside their milking parlor (Figure 9). Previously, they had relied on an electric boiler connected to the SEN, but the solar heater now provided water up to 90°C directly from the abundant sunlight on the farm. They used the hot water to wash the milking machine and to clean the udder cloths. Researchers working for CITMA, the Ministry of Science, Technology, and Environment, reported that hot water primarily was used during peak hours in the SEN. Solar heaters therefore had the potential to drastically reduce the load on the electricity system at this time.[56]

In Pinar del Río's municipal and provincial assemblies, the deputies discussed the introduction of animal traction for waste collection as well as passenger and goods transport. The scarcity of transport fuels undermined the spatially embedded patterns of long-distance and intra-urban transport that had developed as part of the socialist project. This was especially evident in the case of urban waste management when the sidewalks filled up with garbage. Offering political guidance, the provincial politburo announced that Pinar del Río's companies and farms should "declare themselves free from machinery." They commended the work of the Tobacco Company in Consolación del Sur that had achieved "real possibilities to soon be able to declare themselves free from tractors."[57] The Provincial Assembly agreed that the use of animal traction should "continue as a prioritized activity, besides being an essential element of the programa energético."[58]

In stark contrast to the notion of electrification, mechanization, and automation as the techno-material base of a socialist economy, the provincial government portrayed the transition from mechanical to animal traction as a significant accomplishment. In May 1993, the top cadres in Pinar del Río announced that the three thousand work animals that existed in the province in 1990 "among other achievements" had increased to seventeen thousand.[59] Delegates from Pinar del Río later shared this accomplishment in the National Assembly for it to be "generalized" nationally.[60] With some hindsight, Arcadio Ríos and Félix Ponce estimated that nine thousand tractors and five hundred thousand oxen existed in Cuba in 1960. By 1990, the tractors had increased to eighty-five thousand and the oxen decreased to one hundred sixty-three thousand. In 1997, seven years into the special period, the number of tractors totaled seventy-three thousand while the oxen numbered four hundred thousand.[61]

From the beginning of the Revolution, a low degree of motorization characterized the Cuban transport system compared to other countries in the region. This was particularly the case in terms of privately owned cars, which before the special period came in two types: an American import bought before 1959 or a Soviet make, such as a Lada, Moskvitch, or Volga, which was sold at a discount to vanguard workers as part of the *emulación socialista* and to veterans of internationalist missions. According to one estimate there were 32 motor vehicles per one thousand Cubans in 1997, which compared to 45 in Ecuador and 110 in Chile. The low levels of private car ownership may have helped ease the impact of the fuel shortages in the special period, but the energy crisis still radically reconfigured the established patterns of mobility. According to official figures, the average annual distance traveled in publicly owned vehicles decreased from a high of 3,000 km per person in 1986 to a low of 744 km in 1995.[62] Many studies of the special period have also noted that the government imported large numbers of bicycles from China to facilitate urban transport, finally setting up a jointly owned bicycle factory in Havana. Ariana Hernández-Reguant, for instance, reports that about a quarter of a million bicycles were distributed in Havana alone in 1991.[63] During my fieldwork, walking, traveling by horse cart or *bicitaxi* (bicycle rickshaw) were still common—if not the most common—modes of intra-urban transport in Pinar del Río.

In yet another case of energy system redesign, the government promoted the use of biodigesters (*biodigestores*) as a technique for producing biogas. At a piggery north of Viñales, the biodigester consisted of a large black rubber bladder partly dug into the ground. Pig excrement was shoveled into the bladder with water before the bladder was hermeneutically sealed. When the sun shone on the installation, the feces decomposed anaerobically, turning it into methane and a dry residue that was used as a biofertilizer. Company management piped the biogas to fifteen households neighboring the farm to be used as a cooking fuel, free of charge. In the national context, the government had taken an interest in biogas already at the Energy Forum in 1984, but biogas use was still marginal in the early 1990s. In the wake of the Programa Energético, a working group was set up in Santa Clara to promote the installation of biodigesters nationally.[64]

Beside their solar heater, Carlos and Gladys had a biodigester made from concrete on their farm (Figure 10). A study on biogas production at EPG "Camilo Cienfuegos" noted that there was no shortage of organic residue in the livestock company.[65] During my visit, Carlos demonstrated the biogas production process; Gladys then took over to show how the gas entered directly into the kitchen and the cook stove (Figure 11). She found the gas more convenient to use than the rationed kerosene she had used previously. The gas was easier to handle, had no smell, and burned evenly. Carlos and Gladys also used methane from the biodigester in a small

lamp, which greatly reduced their electricity expenses. Nine months before I met them, hurricane Gustav had ravaged Pinar del Río, a Category 4
hurricane moving with wind speeds of up to 240 km/h, toppling pylons
and power lines along its way. In comparison to households that relied on
the centralized infrastructures of the state alone, Carlos and Gladys explained that the biodigester had provided them both with light after dark
and a flame for cooking.

TWO POLITICAL GEOGRAPHIES
OF ENERGY INFRASTRUCTURE

The reinvented energy systems allowed Cubans to carry out tasks that
previously had required electricity or fossil fuels. Carlos and Gladys harnessed the thermal energy they needed to heat water directly from sunlight on the farm. They sourced energy directly in the local context instead
of importing it from elsewhere. At the center of this re-spatialization of
energy use was a reconfiguration of the ecological conditions of sociocultural practice: a shortened chain of energy conversions reenabled
everyday life when the fossil-fuel-based energy infrastructures failed.
Prior to the special period, Carlos and Gladys had relied on a long chain
of energy conversions to heat water. If only the conversions taking place
inside a thermoelectric power plant are taken into consideration, fuel oil
was turned into heat, putting steam in motion, generating mechanical energy in a turbine, and then electricity in a generator. In a transformer station, the voltage would then be stepped up for transmission before it was
stepped down for distribution in another station, and the electricity was
again converted into heat in Carlos and Gladys's water boiler. The water
could then be used to wash their milking machine. Following the second
law of thermodynamics, entropy increased with each energy conversion,
which meant that much more energy had to enter the system as a whole
than the amount of free energy ultimately present in the water circulating
in the milking machine's glass tube.

    With the use of a solar heater, the number of energy conversions was
drastically reduced. Solar energy entered the heater, where it was converted into heat and transferred to the water passing through the solar

Figure 10. Biodigester with "fixed cupola." Photo by Gustav Cederlöf.

heater's pipe. It could then be used to clean the milk tube. As a result, significantly less low-entropy energy had to enter the system as a whole to achieve the same outcome compared to the system that relied on centralized electricity infrastructure. This reconfiguration of the energy system did not increase the economic efficiency of one or several conversions within the system, like the *moñito* or the suspended kerosene bottle did. Instead, the energy system became thermodynamically more efficient in a systemic sense. With fewer energy conversions leading up to the moment of end use, Carlos and Gladys reduced the need for low-entropy input into the energy system as a whole. With the reinvented systems, therefore, diversification did not imply an increasing variety of energy sources incorporated in one system but a variety of systems as such. Systemic energy efficiency entailed reduced energy demand, which was achieved by reconstructing the energy system in its entirety.

If we compare the revived and the reinvented infrastructures more broadly, the latter had radically different physical geographies compared to the state's centralized infrastructures. The construction of the SEN

Figure 11. Methane cook stove connected to a biodigester. Photo by Gustav Cederlöf.

reconfigured the geography of everyday energy use by displacing electricity generation, primary energy extraction, and the ensuing environmental impact from the energy user's space of experience. This kind of energy
infrastructure depended on long chains of spatially and temporally distributed energy conversions. With the reinvented systems, Cubans again
shortened the conversion chains, emplacing the energy sources they relied on in their local environment. Generation and application were again
"fused in the same site," as Michael Anusas and Tim Ingold would argue.[66]
Shorter transmission distances themselves contributed to higher systemic
efficiency. A key purpose of the SEN, however, was to facilitate centralized
redistribution, attesting to a state socialist conception of energy justice.
The SEN integrated the nation territorially and enhanced the infrastructural power of the state, enabling a centrally coordinated energy supply.
In comparison, the reinvented energy systems were place specific; they
were only "generalizable" by taking the socio-ecological contexts in which
they were created into consideration. The physical geographies of the reinvented infrastructures thus challenged the political economic rationale
embedded in the socialist state's infrastructural form. With their biodigester, Carlos and Gladys's supply of cooking fuel only relied on the relations between (mainly) Carlos's work, the biodigester, their defecating
cattle, the sun, and the rubber tube that led methane into the cook stove.
The biodigester, just like the other reinvented systems, provided the conditions of possibility for increased self-sufficiency, allowing its users to
withdraw from the state when it served their interest.

In her study of state power in Mexico, the geographer Katie Meehan
identifies a similar dynamic between the use of materially distinct water
infrastructures. The construction of centralized water infrastructure and
the rapid growth of Tijuana in the 1960s–80s worked in tandem to increase household dependence on a distant, state engineered infrastructure. At the same time, the sustained household use of buckets and barrels
to collect rainwater permitted urban residents to organize their water use
in autonomous spaces, free from state control. Infrastructures with different material qualities were "power brokers" that enabled and circum-
scribed state power. Meehan carefully builds a relational concept of the
state and argues that power springs from "autonomous and irreducible"
objects, such as buckets and barrels, with an inert ability to catalyze social

change. Thus, the contours of the state emerge through human interaction with an infrastructured environment. Somewhat strangely, however, the state is the only "thing" that escapes reification in Meehan's account.[67] In Cuba's special period, opposing political geographical forces of redistribution and autonomy worked through energy infrastructure to tie everyday energy use both more firmly toward and away from a political economic center.

The anthropologist Michael Dove and the physicist Daniel Kammen call technologies such as Carlos and Gladys's biodigester "mundane science."[68] They argue that a systematic bias traverses Western research and development praxis that values high-tech solutions above solutions that are mundane, low-tech, and based on folk knowledge. Dove and Kammen examine solar ovens and passive lights as two examples of undervalued but potentially successful mundane technologies. The former are ovens directly using thermal energy for heating while the latter are plastic bottles filled with water and bleach that provide 60 W of light if fitted into a roof. In contrast to more complex systems, these energy technologies share the physical characteristics of the Cuban wind-water pumps, solar-water heaters, and gravity-irrigation systems. However, from the perspective of mainstream research and development praxis, they tend to be regarded as politically and economically unproductive solutions. They reduce central control, neglect areas for investment and profit-making, and undermine the livelihoods of distant scientists and technicians. From a Marxist-Leninist point of view, they also go against the historically determined development of the productive forces. To the contrary, Dove and Kammen find it remarkable that such mundane solutions would be considered unproductive, or even belonging to a prior "stage" of cultural and social development, when they "upon close study reveal themselves to be elegant solutions to the livelihood challenges of particular socio-ecological times and places."[69]

The re-spatialization inherent in the Cuban reinvented systems lends itself to a closer comparison with another form of mundane science examined by Dove and Kammen. Swidden solves a basic dilemma in agricultural systems when energy and nutrients are extracted from the soil at the time of harvesting. Harvesting leaves the agroecosystem in need of low-entropy input in the form of organic materials and fertilizers to remain productive.[70] Using swidden techniques, farmers solve this issue by leaving land fallow for a period that is longer than the period of cultivation. The biomass that regrows is then burned, which returns nutrients and energy to the soil. Industrial agriculture, by contrast, solves the dilemma with inputs of synthetic fertilizers, which are produced using large quantities of natural gas.[71] While industrial agriculture gives unparalleled returns to farm area, Dove and Kammen argue that swidden gives unparalleled returns to labor. And yet swidden is frequently misrepresented as archaic and unproductive when compared with scientific, high-yielding industrial farming.[72]

Swidden and industrial agriculture are distinguished not only by their exchanges with the natural environment but also with the social environment. Dove and Kammen argue that Indonesian swidden systems were sustainable not only in an ecological sense, but also because they were "local and social in character." Swidden was sensitive to context-specific ecologies and was based on reciprocal social relations that safeguarded farmers in the case of crop failure. By contrast, industrial systems were "distant and economic in character," making use of energy purchased with money gained from market trade.[73] Hence, industrial and nonindustrial, modern and mundane technologies distinguish themselves by their relations to an extra-local political economic center. Dove and Kammen point out that the use of mundane techniques does not imply a "lack" of development or modernity but rather allows people to distance themselves from such globalizing, disempowering projects. "It is not isolation from political-economic centers that forces people to practice swidden cultivation," they write, "it is swidden cultivation that permits people to distance themselves from such centers."[74]

In comparison, Carlos and Gladys's biodigester solved both a basic dilemma of energy provisioning and weakened their ties to an extra-local political economic center. These social relations were conditioned by the materiality of the energy system, but they were not determined by it. In the Cuban revolutionary narrative, the industrial SEN was regarded as a precondition for socialist redistribution, and indeed, the national grid co-constituted the socialist state and embodied a qualitatively different—but not necessarily less desirable—political economic rationale than many of the reinvented energy systems. As place-bound solutions increasing local autonomy, the reinvented systems were not available to all, and they pro-liferated as a result of the disruption of centralized supplies, which had so-cially harmful consequences. In other words, within a framework pitting degrowth and eco-socialist ideals against each other, different spatially constituted imperatives of development came into conflict in the process of infrastructural change.

The question of whether particular technologies also have particular political qualities is of course the subject of a long and contested debate. In chapter 1, we saw how Che Guevara argued that the development of the productive forces led to a massive concentration of productive power in particular locations, but also how this technological capacity became in-creasingly interconnected. The form of such industrial infrastructure cre-ated the enabling conditions for centralized planning and political control on a national scale. In his essay "On Authority," Friedrich Engels simi-larly argues that the large-scale infrastructure he witnessed in Victorian Britain imposed a need for human coordination in time and space. The combined and interdependent work tasks involved in the operation and maintenance of automated machinery demanded human subordination to a hierarchical system of social organization. "Wanting to abolish au-thority in large-scale industry," he writes, "is tantamount to wanting to abolish industry itself, to destroy the power loom in order to return to the spinning wheel."[75] While such authority was oppressive under capitalism, it was also the wellspring of socialist development: highly developed pro-ductive forces enabled collective ownership and democratic control over the means of production.

At the other end of the socialist spectrum, many advocates of small-scale solar development argue that solar technologies, technically speak-ing, are more egalitarian than large-scale infrastructures. Hermann Scheer, who as a member of the German Social Democratic party invented the "feed-in tariff" to incentivize increasing solar energy use, argues that renewable energy technologies not only are more comprehensible and accessible for nonexperts but also enable a greater degree of local self-determination. When energy is transmitted to the end user through extra-local infrastructure, it makes the consumer dependent on physical and social processes operating on a larger, more abstract scale, "subordinat-ing" local producers and consumers to the logic of grid-scale operators.[76]

By contrast, the short conversion chains of renewable energy technologies physically prevent large corporations from monopolizing key energy flows. Short conversion chains, he argues, "make it impossible to dominate entire economies."[77] The Cuban energy expert Luis Bérriz makes a similar argument. Bérriz is the president of CubaSolar, a society established at the height of the special period to lead a transition to renewable energy use in Cuba. In a series of articles in the journal *Energía y tú*, he argues that the dispersed quality of solar energy makes it a weapon of the people against capitalism and imperialism. The material forms of renewable energy technologies prevent the rich from controlling them: "The Sun cannot be blockaded, it cannot be dominated, it cannot be destroyed. Solar energy is a weapon of the peoples, of socialism, and it is the only [energy source] that can produce the true kind of economic and social development that humanity needs."[78]

The efforts to develop both centralized and decentralized energy infrastructures in the special period were situated between these two broader arguments for socialist development. Indeed, the tension between processes of centralization and decentralization indicates that energy transitions are moments at which key political questions about the spatial patterns of energy use are negotiated. It also indicates that decentralized energy infrastructures in and of themselves have no greater democratic potential than large-scale, centralized infrastructures. "The point, rather," as Timothy Mitchell argues, "is that in battles over the shape of future energy systems the possibilities for democracy is at stake."[79] Writing in a different context, Doreen Massey has warned against arguments that succumb to a spatial fetishism and attribute political content to particular geographical forms. "It is not spatial form in itself but the particularities of the social construction of that form in any specific instance, that should be the focus of political evaluation," she writes.[80] In Cuba, the SEN embodied a scalar strategy that enabled energy redistribution over a large territorial area in the pursuit of justice and socialism. Infrastructure diminished social difference across space. The energy systems invented in the special period instead increased local autonomy at a time when the island's geopolitical relations were severed and its centralized infrastructures failing. The new energy systems bypassed the socialist state, allowing people to reduce their dependence on it—indeed withdraw from it—in order to reconfigure the infrastructural form of energy use. In each context, we must ask: which was the most progressive political goal?

THE LOGIC OF DEINDUSTRIALIZED SOCIALISM

From its headquarters on Novy Arbat in central Moscow—a modernist skyscraper overlooking the Moskva River—the CMEA bound the socialist bloc together, coordinating trade between the socialist states. To celebrate thirty years of socialist cooperation and development, the 1979 edition of CMEA's *Statistical Yearbook* reported that the total rated horsepower of tractor engines in Cuba had increased by almost 50 percent in the course of the 1970s.[81] This statistic served as an index of socialist development, seeing that in the industrial socialist future mechanical horsepower replaced organic muscle power, emancipating the working class from labor. In hindsight, it seems like a paradox, then, that Cuban Communist Party officials and members of the Poder Popular actively encouraged the production of energy infrastructures that bypassed the state in the special period. It seems even more paradoxical that they portrayed these systems as revolutionary achievements—as outcomes of struggle and popular resourcefulness. Indeed, many of the reinvented infrastructures reverted everyday practices of energy use to infrastructural configurations in use prior to the Revolution. Oxen replaced tractors for traction, gravity replaced electricity for irrigation, and methane replaced kerosene for cooking. Without a steady supply of oil, the productive forces de-developed.

In the revolutionary narrative, however, this was not a contradiction. As a category of experience, the "special period" both explained and legitimized the development of nonindustrial energy systems. During Cuba's interim epoch, the reinvented energy systems were "special measures" in a "special period." It was a historical necessity for the Cuban people to employ oxen and horses instead of tractors and cane harvesters for the Revolution to survive. The reinvented energy systems signified Cuban "resistance" to reactionary forces. But there was also an economic logic acting to overcome the implied contradiction. This logic was rooted in the distinction between use value and exchange value and the imperative to de-commodify resources in the socialist state. Carlos and Gladys's biodigester only provided biogas to their family, a fact that arguably gave them access to an undue amount of cooking fuel compared to the national population at large, especially in a time of crisis. However, as they did not sell the gas for a profit, the biodigester only provided them with a use value. Similarly, the biogas produced in the Viñales piggery was distributed to fifteen households neighboring the farm. But seeing that the methane was distributed without counterclaim, it remained uncommodified in the eyes of the socialist state, even as the potential to commodify the biogas existed in the infrastructural system, with its physical connections between central biodigester and cook stoves. As long as the reinvented systems provided use values alone, they indirectly helped the government reduce the demand for centralized energy supplies while maintaining its hegemony over resource exchange. The otherwise resolutely centralized state therefore benefited from such decentralized infrastructural solutions.

Meanwhile, the government looked with concern on the informal trade of charcoal. If a peasant had produced charcoal for self-consumption, and perhaps shared surplus freely with his or her neighbors, the government would likely also have looked favorably on this practice. But the materiality of charcoal production, paired with the regulation of land use and the urban demand for cooking fuels, created opportunities for the peasantry to commodify this energy form—to trade it as an exchange value. Charcoal therefore conceptually challenged the socialist state's distributional monopoly and vision of energy justice. As we have seen, Party officials also portrayed the informal trade of charcoal as opportunistic—as "abusive" and "unpopular"—even as they allowed it to continue. Nevertheless, when a lack of capital and an inherently oil-dependent industry impeded the government's attempts to establish a nationally territorialized energy supply system, informal trade reduced the friction between state and citizen over basic energy needs. In the struggle against imperialism, a pragmatic approach and a temporarily more porous state formation was needed.

In the early 2000s, however, things were about to change again.

5   The Energy Revolution

On May 5, 2004, the thermoelectric power plant "Antonio Guiteras" in Matanzas was scheduled for twenty-nine days of maintenance. The Guiteras was one of the largest plants in the country and contributed 15 percent of the SEN's total capacity. At 3 a.m., the night shift started cooling the turbine to shut down the plant, but mid-process they failed to close the valves letting out steam, and the temperature fell too quickly. The rapid temperature change deformed the turbine shaft and left the plant inoperable. A Ministry of the Interior investigation concluded that the incident was unintentional and did not constitute an act of counterrevolutionary sabotage. The turbine was shipped to a workshop overseas (the location was left undisclosed to prevent US-sponsored interference) and the repair works dragged on from early May to the end of September.[1]

The Guiteras breakdown had far-reaching consequences. To describe an electricity system, engineers distinguish between its nameplate capacity and availability factor. The nameplate capacity is the maximum power the generators in a system can achieve, while the availability factor designates the length of time the power plants can generate electricity in a given time period. In the 1980s, the nameplate capacity of the SEN increased with the construction of new power plants, testifying to socialist progress, and the system had an availability factor of around 80 percent. The availability factor then dropped to 50 percent during the special period due to the fuel shortages, leading to power cuts, before it rose to almost 70 percent following the retrofitting works in 1997 when the thermoelectric plants were adapted to handle Cuba's sour crude oil. After the Guiteras breakdown, it fell back to 57 percent.[2]

The issue extended beyond the turbine shaft in Matanzas. Though the power plants now ran on Cuban crude, the use of such a corrosive fuel meant that they still required more frequent maintenance compared to when they ran on "sweeter" Soviet fuel oil. When the "Antonio Guiteras" was forced offline, UNE had to put maintenance planned for other power plants on hold so that these plants could compensate for the loss of the Guiteras. The situation was complicated by the thermoelectric plant "Lidio Ramón Pérez" in Felton that had come offline following a transformer failure in November 2003. The repair works in Felton carried on until early 2004 when UNE had retrofitted a transformer previously installed at the nuclear construction site in Juraguá. Both incidents set a vicious circle of postponed maintenance works in motion that kept the SEN's availability factor critically low long after the "Antonio Guiteras" came back online.

On August 13, matters worsened further. Except for the Storm of the Century in 1993, hurricane Lili in 1996, and hurricane Georges in 1998, Cubans had been relatively spared from hurricane damage since the onslaught of hurricane Kate in 1985. The 2000s turned out to be a different story altogether. In 2002, hurricanes Isidore and Lili battered the island within weeks of each other, before hurricanes Charley and Ivan struck in 2004, Dennis in 2005, followed by Gustav, Ike, and Paloma in 2008. Amid the havoc wreaked by hurricane Charley, eighteen pylons holding 110 kV transmission lines between Mariel and Pinar del Río and eight 220 kV transmission towers between Mariel and Havana were knocked over. The ruptures shortcut the thermoelectric plant "Máximo Gómez" in Mariel from the rest of the already debilitated SEN, and even more drastically, the fallen high-voltage cables left ten municipalities in the capital and the rural province of La Habana, and the entire province of Pinar del Río, without electricity for eleven days.[3]

In Pinar del Río, the population resorted to many of the strategies and informal spheres of exchange they had developed during the special period. The *pinareños* I interviewed a decade later recalled the daily "struggle" of obtaining food, fuel, and water while coping with the intense summer heat. Without electricity, there was no way of storing fresh food. "There was a terrible heat," Vilma said. "You know, there was no cold water and then you had to cook with the *Pike* [kerosene burner] and eat the food straight away, because the refrigerators didn't work." Mélanie Josée Davidson and Catherine Krull argue that women suffered the consequences of the hurricane in particular because of the gendered division of labor in households.[4] Again, the crisis-stricken Cubans resorted to their "inventiveness" to make do. "I remember a neighbor had a car battery," Vilma said; "you know, the Cubans invent a lot—and everyone in the *barrio* came to watch TV."

In the state's rendition of events, emergency quotas of kerosene and LPG were distributed to the population for cooking and indoor lighting, in addition to bread, milk, and soya yoghurt rations. The leaders of the Communist Party in Pinar del Río raised concerns over the informal distribution systems that were appearing when "in situations like these there are unscrupulous persons who go about reselling [goods] for sham prices."[5] The situation was particularly worrisome in the municipal water systems where the pumps stopped running without electricity. The government scrambled portable diesel generators to supply emergency power, and *Guerrillero* reported that eleven hand-pumped wells, dug as a war resource, were activated in Pinar del Río city.[6] A brigade of electricians was called out to redraw power lines from the "José Martí" sugar mill to the municipal pumping stations in San Cristóbal and Candelaria.[7]

In late September 2004, when hurricane Ivan also had swept across the island, Fidel Castro staged a public event to "inform and also educate the population" on the situation.[8] Over the course of three evenings, he participated in *Mesa Redonda* (Round table), a nightly TV show, taking on the role of the people's advocate to interrogate ministers and top managers in the electricity sector about the overwhelming problems. Soon after, he announced that the country would embark on a nationwide campaign that would radically change the way energy was used. The campaign would achieve three things: first, stability would be brought back to everyday energy use; second, the SEN would become more resilient to hurricanes; and third, in the process, large oil "savings" would be made. The National

Assembly declared that 2006 would be the "Year of the Energy Revolution," and at the end of the campaign, the carbon intensity of the Cuban economy would have decreased by a remarkable 32 percent and its energy intensity by 44 percent.[9]

FROM TEN TO A THOUSAND: RE-SPATIALIZING
THE NATIONAL ELECTRICITY SYSTEM

One of the most significant aspects of the Energy Revolution was a radical overhaul of the SEN. The government decided to reconfigure the national grid in three ways: geographically, ecologically, and by undertaking long overdue repairs. Since the late 1980s, the SEN had expanded from integrating five to eight large oil-fired thermoelectric plants, and as noted, a plant such as the "Antonio Guiteras" concentrated 15 percent of the national generating capacity in one location. Following the events of 2004, the government reasoned that this geographical form made the infrastructural system vulnerable. "This was madness," Fidel Castro remarked on one occasion, referring to the construction of the thermoelectric plants; "we must have been full of dogmatism and schematism."[10] The government instead proposed to deconcentrate the generating capacity into a system with generators geographically spread out in their thousands. This was "a paradigm shift" in the country's energy planning, a textbook for Cuban engineering postgraduates noted.[11]

The distributed system worked in a twofold manner. First, UNE installed diesel and fuel-oil generators in "emplacements" (*emplazamientos*) all over the country (Figure 12). By the end of 2006, state media reported that emplacements of 20–40 diesel generators were up and running in 116 of Cuba's 169 municipalities. Together with almost 700 fuel-oil generators, their capacity added up to 1.3 GW out of a total 4.7 GW in the SEN. In 2009, the distributed capacity surpassed 2.0 GW.[12] In anticipation of the installations, the magazine *Bohemia* announced that the new generators would be "the spine of the new energy system."[13] Luis, who supervised the installations nationally, explained that as soon as a voltage drop was registered in one area, the distributed system automatically fired up a number of generators nearby to make up for the power deficiency. Compared

Figure 12. Diesel generator emplacement in central Pinar del Río. The emplacements consisted of fuel-oil and diesel generators kept in blue freight containers, each with a protruding smokestack. At the emplacement in the picture, which was wedged between the railroad, the city's ice factory, and a colonial-era residential area, there were twenty diesel generators. At a larger emplacement on the edge of the city, there were thirty diesel and three fuel-oil generators. Diesel was used when there was a local shortage, "when there is a higher demand, or if there is a cyclone or something." When there was a larger power deficit in the grid, especially during the demand peak in Havana, the fuel-oil generators were fired up. Photo by Gustav Cederlöf.

to the thermoelectric power plants, the generators had a very high availability factor, reportedly being available for production 90 percent of the time.[14] While keeping the grid nationally integrated, the generators in emplacement allowed the energy system to operate across independent smaller-scale sections.

Second, UNE installed emergency generators in places deemed socially important by the government. Emergency generators existed in hospitals already prior to the Energy Revolution but were also installed in pharmacies and nursing homes, in bakeries, media outlets, aqueducts, meteorological stations, and schools for practitioners in the Cuban-Venezuelan medical program Misión Milagro (Mission Miracle) among other places.

The diesel generators could supply electricity in isolation in case of grid failure, complementing the SEN's networked spatiality with a point-like configuration. In his May Day speech in 2006, Fidel Castro reported that 2,755 emergency generators were ready for use, together accounting for 296 MW.[15] Production reports from 2009 indicate that distributed generators of both kinds generated 25 percent of all electricity that year.[16]

With the distributed generators working alongside the thermoelectric plants, oil was still the dominant fuel in the SEN. But for a second time following the Programa Energético, the government attempted to transform the energy system ecologically, territorializing the supply of primary energy on the island. Publications stemming from the Energy Revolution strongly focus on renewable energy sources and a diversification of the energy mix. For example, the textbook for engineering students explained that renewables were a natural part of the new paradigm: "Distributed generation is the perfect framework for using renewable energy sources, [a wide geographical distribution] being their principal characteristic."[17] Again, waste biomass from the sugar industry was identified as an important resource, and the government also singled out wind as one of the most promising renewable energy sources. It assigned the Institute of Meteorology to map wind speeds in thirty-two zones across the country, and in four of these, the wind potential was deemed sufficient for electricity production. Between 2007 and 2010, three wind parks opened on Isla de la Juventud and in Gibara, Holguín, with a total capacity of 11.7 MW.[18]

The integrated network of smaller-scale diesel and fuel oil generators was a unique feature of the Cuban electrical industry, but the interest in renewables tied into a broader pattern of industrial change in the Caribbean. Several Caribbean island-states developed policies for investments in renewables in the late 2000s, aiming to reduce carbon emissions but, perhaps more significantly, to protect themselves from volatile oil markets. While the specific circumstances vary from state to state, recent studies show that questions of land acquisition and the regulation of a more diverse supplier market presented challenges to the success of these policies, but these did not surface in Cuba, where there was no market for land and no market for energy supply.[19] Even so, government figures indicate that renewable energy use increased only marginally during the Energy Revolution (Figure 13). The wind parks made a modest contribution

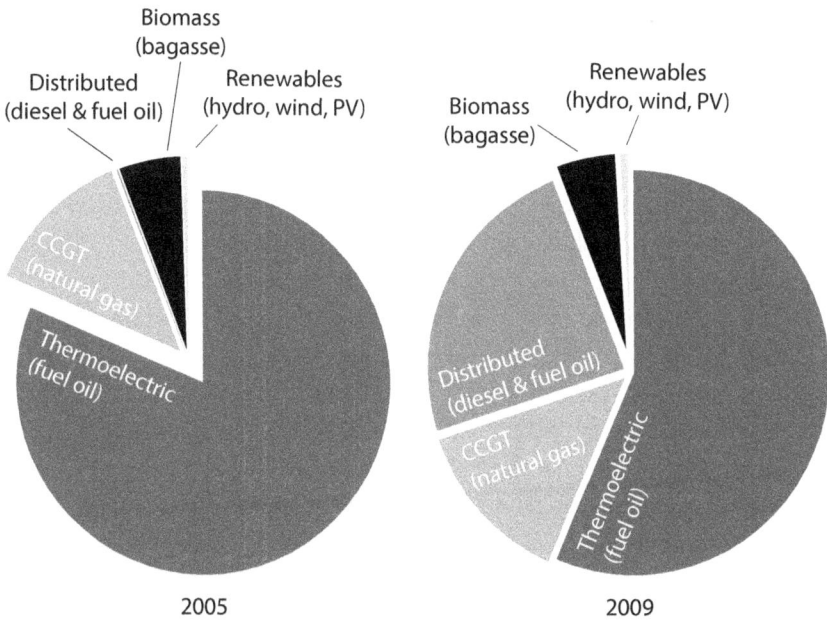

Figure 13. Electricity output in the SEN per generating technology before and after the Energy Revolution. *Source:* ONEI, *Anuario estadístico de Cuba 2016*, table 10.6.

to the SEN, together contributing no more than about 4 percent of the "Antonio Guiteras's" total capacity. Production from bagasse and cane leaf, though strategically important, decreased throughout the 2000s following the dismantling of the sugar industry. Hydroelectric output increased slightly but with a large annual variation, and while generation from wind and solar increased from 2008, it accounted for only 0.2 percent of total output in 2014. At the same time, the system of distributed diesel and fuel-oil generation contributed 19.9 percent of the total; Energas's three CCGT plants stood for 14.4 percent; and the thermoelectric plants contributed 60.6 percent.[20] Thus, while the energy mix was diversified, the SEN remained dependent on fossil fuels.

Third, UNE upgraded the transmission and distribution network, which it had failed to maintain for many years due to a lack of spare parts. Worn transformers and cables created great heat losses in transmission, and in the early 2000s, close to one-fifth of all electricity reportedly dissipated as heat in the system.[21] In coastal areas, the salty sea winds oxidized the SEN's aluminum cables, creating a fine white rust that conducted electricity poorly.[22] The utility also identified fifteen thousand distribution zones with low voltages due to poor wiring. According to Carlos Lage, the executive secretary of the Council of Ministers and the de facto prime minister, nine thousand low-voltage zones had been attended to by June 2007. Linesmen had also replaced 67,319 posts, installed 11,700 new transformers, and made 357,335 new cable connections.[23] In May 2006, Fidel Castro announced that transmission losses had decreased from 18 to 15.75 percent and that losses would add up to around 11 percent once the upgrades were complete.[24] These figures, while difficult to verify, clearly indicate that the maintenance work aimed to reduce heat losses, allowing less oil to be combusted in the thermoelectric plants.

The overhaul of the SEN had three outcomes that merit attention. First, the electricity supply was properly stabilized for the first time since the 1980s, putting an end to the frequent blackouts. "I will limit myself to saying that Pinar del Río won't know of the *apagones* [blackouts] anymore," Fidel Castro proclaimed to ovations when he inaugurated the country's first emplacement in the province.[25] Blackouts could occur temporarily if a tree fell over a line, a transformer burned, or a hurricane required supply to be cut off, but there would be no lack for a shortage of generating capacity. Norberto, an electrician with the Pinar del Río branch of UNE, explained that now that there were generator emplacements in practically all municipalities, there was no longer a shortage of power. The magazine *Bohemia* boasted that already in March 2006 the new generators had compensated for the thermoelectric plants "Antonio Guiteras" and "Carlos Manuel de Céspedes" when these had come offline.[26] Everyone I spoke to on the subject could testify to this: the Energy Revolution had brought stability back to the grid.

Second, the Energy Revolution decarbonized the electrical industry in relative terms. Despite lacking the economies of scale of the large thermoelectric plants, the distributed generators were more efficient in terms of converting oil into electricity. Under peak conditions, the new fuel-oil generators used less than 210 g/kWh, in contrast to the voracious thermoelectric plants from the 1960s, 1970s, and 1980s—"the 'Guiteras,' or those monstrous plants in Santiago de Cuba . . . using between 300 and

350 grams of fuel oil per kilowatt of electricity," as Fidel Castro noted.[27] Natural gas and some renewables also displaced fuel oil in the energy mix. Critically, the distributed generator system reduced the need for energy transmission across the entirety of the dilapidated national grid, increasing the SEN's systemic energy efficiency. With generators located in just short of all municipalities, the electric current only had to travel short distances to reach the place of consumption, reducing heat losses.[28] Not least, the upgraded grid reduced heat losses in transmission overall as new cables and transformers caused less resistance to the current.

Finally, with the deconcentration of generation, the SEN became more resilient to hurricane damage. Physically, control over the electricity supply was re-scaled to smaller territorial units when municipally employed electricians, working in emplacements, controlled locally available generators with a high availability factor. In January 2006, Pinar del Río became the first province with enough generating capacity to be electrically self-sustaining.[29] Carlos Lage confirmed that by the end of the Energy Revolution, all provinces would have a greater capacity than provincial consumption demanded. If a province would be cut off from the grid in the way Pinar del Río had been following hurricane Charley—or following a military invasion for that matter—supply could now be restored from within an isolated section of the grid.[30] The benefits of the SEN's deconcentrated spatiality were evident already in 2008 when hurricane Gustav hit Pinar del Río. Despite the most severe material damages ever following a hurricane in Cuba up to that point, and a general blackout in the municipality of Viñales that lasted thirty-two days, vital social services had continuous electricity supply with access to emergency generators and isolated grid-sections, as did bakeries and urban water infrastructures.[31]

The deconcentrated and decarbonized energy system notwithstanding, the SEN remained a heavily centralized "national" infrastructure. Electricians were still employed by the state utility, which was organized hierarchically across Cuba's fifteen provinces and 169 municipalities under the aegis of the Ministry of Heavy Industry. To run the generators at their disposal, they relied on centralized deliveries of fuel and diesel oil, which were administered by Cupet via Cuba's four refineries. In a portrait from the "Ciego Centro" emplacement in Ciego de Ávila, *Bohemia* explained that if the generators ran at their most efficient level, they consumed 53 m³ of fuel oil per day. Cupet workers delivered the oil from the "Sergio Soto" refinery in Cabaiguán, and this was no different from the fuel-oil supply to the large thermoelectric plant "Diez de Octubre" in Nuevitas.[32] Control over the electricity supply thus remained firmly vested in the socialist state. Indeed, as Jessica Barnes has argued in the context of infrastructural maintenance on the Nile: "What is being maintained . . . is not just the infrastructure, but the state's control over that infrastructure."[33] The state had materially decentralized its infrastructural form but reinforced the centralizing institutional logic of it.

THE BOLIVARIAN ALTERNATIVE

In parallel to the operations on the national scale, transformative change took place in Cuba's international relations. Energy flows in and out of the SEN connected to a new regional energy regime. From the jetty at Hotel Jagua in Cienfuegos, the trappings of the Cold War geopolitical order were in plain sight. Cuba's abandoned nuclear reactor made itself present to the south, while the towers of the formerly Soviet-sponsored "Camilo Cienfuegos" oil refinery and the "Carlos Manuel de Céspedes" thermoelectric plant broke the sky further to the north. The heavy industry formed a counterpoint to the hummingbirds swooping through the air. In 2007, Hotel Jagua played host to the Fourth Summit of PetroCaribe: a circum-Caribbean trade alliance again providing Cuba with oil imports. A photo exhibition at the hotel celebrated the summit, capturing a beaming Raúl Castro alongside the alliance's benefactor, Venezuela's president Hugo Chávez.

In the oil frenzy of the 1950s, oil was identified both as the source of Venezuela's underdevelopment and the promise of its development. The principal concern for General Marcos Pérez Jiménez and the elected governments that succeeded his dictatorship was to secure access to export markets in order to pay for often grandiose development projects at home. "Based on the expansion of the state's oil rents," Fernando Coronil writes of Venezuela's period of rentier liberalism, "each person's interests came to depend on the realization of the nation's own ends."[34] In the early 1980s, the Venezuelan economy took a direct hit in the Latin American debt crisis, and President Andrés Pérez agreed to a loan from the International Monetary Fund contingent on the implementation of neoliberal shock therapy. Riding on a wave of social protest against the austerity and the corruption that followed, Hugo Chávez was elected as president in 1998 promising to increase public participation in political and economic life and to invest oil revenue in social programs targeted at the poor, so-called *misiones*.[35]

Chávez linked his national program to a distinctly geopolitical agenda. From the outset, he took an active stance in the Organization of the Petroleum Exporting Countries (OPEC) seeking to drive up oil prices internationally. Meantime, placing himself in the lineage of Simon Bolívar, he worked to create a Latin American union in which Venezuela would tender oil and financial capital on beneficial terms to countries sharing Venezuela's experience of colonial and neocolonial exploitation. The maneuver was designed to undermine the free market, guaranteeing oil supplies to economically vulnerable Latin American nations while making high-income countries pay more for their consumption.[36] For Chávez, Cuba's internationalist history, extending from its military interventions in southern Africa to the medical support it offered Venezuela after the calamitous Vargas landslides, made the socialist state a key partner, giving political clout to Chávez's anti-imperialist project.[37] In October 2000, Fidel Castro and Hugo Chávez formalized the partnership and started to integrate the Cuban and Venezuelan national economies through a Comprehensive Cooperation Agreement. Soon after, the agreement was amplified in a regional free trade organization christened ALBA—the Bolivarian Alternative for the Peoples of Our Americas—*alba* also meaning dawn in Spanish.[38]

ALBA granted Cuba access to Venezuelan oil shipments on the terms of the Caracas Energy Accord. This treaty concretized Chávez's anti-market strategy so that Venezuela supplied eleven Central American and Caribbean countries with oil on long-term, low-interest conditions. Under the ALBA umbrella, the Accord then transfigured into a new trade organization, PetroCaribe, in 2005. The inaugural members of PetroCaribe framed the alliance in a narrative echoing the Prebisch-Singer thesis. In the capitalist world economy, the PetroCaribe treaty posited, high-income consumer societies were able to waste resources on an enormous scale, and the costs of these excesses were systematically reflected in the price of the hydrocarbons fueling the globalized economy. In the meantime, the peripheral Caribbean island-states remained structurally dependent on oil imports, which they could only finance with income from agricultural exports and tourism. Unequal exchange in the world system thus left the Caribbean countries vulnerable to price fluctuations and inhibited their economic and social development. However, as a southern nation endowed with great oil wealth, Venezuela could offer its regional partners oil on terms that undermined unequal exchange. PetroCaribe, then, produced an anti-imperial infrastructural space, "contributing to the energy security, the socioeconomic development and the integration of the countries of the Caribbean."[39]

Two economic mechanisms regulated trade within PetroCaribe. First, the importing countries could defer payment of a set percentage of the world market price for oil over an extended term. With prices below twenty dollars per barrel, 5 percent could be deferred, after which the fraction increased in eight price bands up to a 50 percent deferral when prices exceeded one hundred dollars. The long term was defined as a period of seventeen to twenty-five years depending on the market price. Rather than the deferred payments being sent to the Venezuelan oil company PDVSA, however, they entered a regional development fund from which the Petro-Caribe member states could apply for financing of social and economic development projects.[40] Second, Venezuela agreed to receive partial payment in the form of a countertrade of goods and services. In the same way that CMEA had legitimized trade with the Soviet bloc's underdeveloped countries, a politically negotiated bartering arrangement in PetroCaribe was seen as a method to undermine the structural inequalities existing between exporting and importing countries. PetroCaribe reports also show that Nicaragua, the Dominican Republic, Guyana, and Jamaica had offset 2.73 billion dollars in 2014 by trading oil for products including rice, coffee, cows, and calves.[41]

The reports do not disclose if or how much Cuba paid for its oil in kind. Yet by 2010, almost thirty thousand Cuban doctors lived and worked in some of the poorest *barrios* in Venezuela, taking part in a number of medical *misiones*. The largest of these was Misión Barrio Adentro (Mission Inside the Neighborhood) through which Cuban doctors served communities in which Venezuela's predominantly white medical staff refused to work. In Misión Milagro (Mission Miracle), Cubans doctors offered Venezuelans eye surgery and Misión Sonrisa (Mission Smile) focused on dental care.[42] While the trade of oil and medical services anchored Cuban-Venezuelan relations, this was not a direct bartering relationship; rather, PetroCaribe and the *misiones* constituted four projects out of hundreds in ALBA. Viewed from the inside, the anti-imperialist logic of PetroCaribe also made it difficult to evaluate the value of trade in financial terms. Charged with the accusation that Venezuela was giving away oil to Cuba, Hugo Chávez argued that Cuba's contribution hardly could be measured in terms of exchange value. Instead, Cuban-Venezuelan trade was predicated on a post-neoliberal political economic logic in which the two partners provided each other with use values for mutual benefit:

> If we start to add up, centavo by centavo, Cuba's contribution is worth 10 times the value of the oil that we send to Cuba. . . . If a country had to contract 30,000 doctors from the United States or Europe to work in the *barrios* and the poorest towns, to live among the indigenous population; build the medical facilities; bring the laboratory equipment and operating room equipment; medicines, how much would a capitalist company or country charge us for that. . . . I don't think that can be reduced to a financial calculation.[43]

To operationalize PetroCaribe, the signatories established the alliance as an organization with its own infrastructure. PDVSA was tasked with forming a subsidiary called PDV Caribe to oversee the logistics, and the member states then agreed to enter into joint ventures with PDV Caribe to develop regionally controlled energy infrastructure within their national jurisdictions. In this economic geography, Cuba became a hub for interregional refining and shipping. PDV Caribe and the Cuban shipping company Internacionales Marítimas formed TransALBA, a firm owning two oil tankers and otherwise contracting Cuban vessels for regional oil shipments.[44] In the early 1990s, Cupet had also been forced to mothball the newly built "Camilo Cienfuegos" refinery in Cienfuegos, lacking crude oil to refine in it. With Venezuelan capital injected into the joint venture Cuvenpetrol S.A., the refinery was renovated to process Venezuelan crude for Cuban consumption and regional distribution. Reinaugurating the refinery during the Fourth Summit of PetroCaribe, Hugo Chávez declared, "Oil,

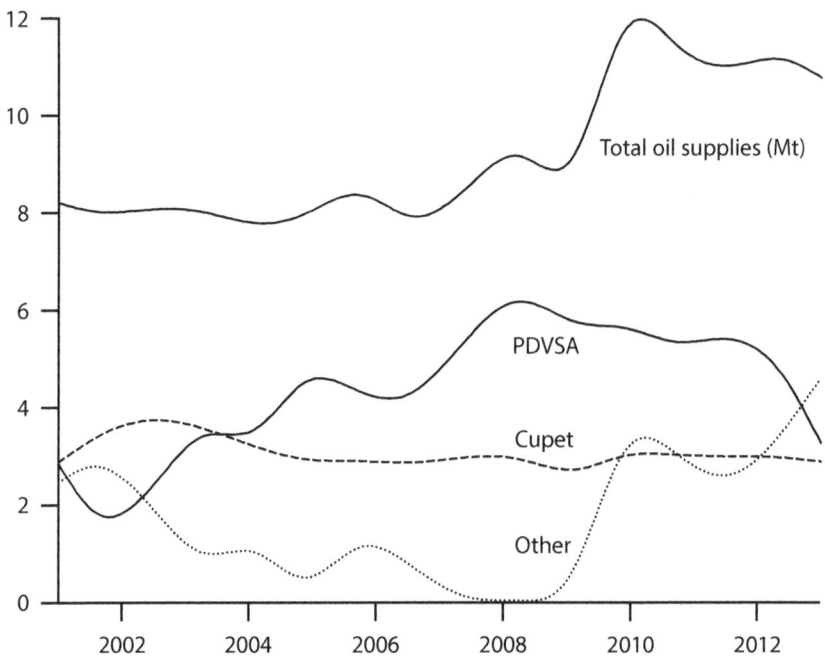

Figure 14. Cuban supplies of crude oil and oil derivatives, 2001–2013, including sources of origin. *Sources:* MPetroMin, *Petróleo y otros datos estadísticos* (51 ed.), table 77; ONEI, *Anuario estadístico de Cuba 2014*, tables 10.4, 10.7.

today, now, is being converted from an instrument of domination into an instrument for the liberation of our peoples through the platform Petro-Caribe."[45] Thus, oil was again flowing into Cuba under an alternative political economic arrangement to the world market norm, rooted in a radical narrative of Latin American liberation.

There are many unsubstantiated figures circulating on the extent of Venezuela's oil exports to Cuba as part of PetroCaribe. The agreement itself does not specify import quotas, and Cuban statistics only report on total imports, failing to distinguish between countries of origin. Reports from the Venezuelan Ministry of Popular Power for Oil and Mines nevertheless indicate that Cuba received 51,720 barrels of crude oil and oil derivatives per day (bpd) in 2001. Imports then peaked at 111,723 bpd in 2008 before falling back to 56,633 bpd in 2014 and even lower levels after this date.[46] In the most generous scenario, this equaled at most 2.8 Mt in

2001, 6.1 Mt in 2008, and 3.1 Mt in 2014, which was still a far cry from the 12 Mt once imported from the Soviet Union.[47] A PetroCaribe management report from 2014 records that Cuba had an import quota of 98,000 bpd (5.3 Mt) that year, which was completed to 74.2 percent (72,716 bpd, or 4.0 Mt).[48] Even so, it is difficult to assess the veracity of statistics like these, not least since Hugo Chávez staffed PDVSA with political allies in the early 2000s, after which at least its financial reports became highly unreliable. Under any circumstances, at the time of the Energy Revolution, oil was again flowing into the Cuban economy and the re-spatialized SEN from a political ally under nonmarket conditions (Figure 14).

REVOLUTIONARY ETHICS AND THE POLITICS OF ENERGY EFFICIENCY

Paradoxically, with oil deliveries secured from Venezuela, allowing Cuban oil consumption to increase in absolute terms, the Energy Revolution was framed as a nationwide effort of *ahorro*. While it had secured access to one energy source through anti-imperialist solidarity, the socialist revolution had also (re)discovered another resource for development: energy efficiency. The public appeal for *ahorro* was taken to an unprecedented level as part of the campaign. Across the island, murals and billboards beseeched the population to *ahorrar* (Figure 15), and Fidel Castro called on all parts of society to foster a rational culture of energy use. Tying together the formation of PetroCaribe on an international level, the reconfiguration of the SEN on a national scale, and everyday energy use in households, *ahorro* became a nodal point in a narrative of revolutionary progress. While state publications identified four concrete reasons behind the need for energy conservation, the effects of the *ahorro* discourse went beyond these explanations. *Ahorro* did political work: firmly rooted in quantitative measurements of energy consumption and energy efficiency, this concept allowed the government to reinvigorate a nationalist imaginary of revolutionary change after the special period.

The first reason behind the call for *ahorro* was economic. According to one source, the cost of the Energy Revolution exceeded two billion dollars, but government representatives reported that the campaign would

Figure 15. "In YOUR consumption, saving counts!" Mural painted by the local branch of UNE in Holguín. Another common slogan appearing on billboards was "*Conjuguemos el verbo* ahorrar. *Yo ahorro, tú ahorras, él ahorra, nosotros ahorramos, ustedes ahorran. Cuba ahorra. Una revolución con energía.*" (Let's conjugate the verb *ahorrar*. I save, you save, he saves, we save, you save. Cuba saves. A revolution with energy.) Photo by Gustav Cederlöf.

largely pay for itself through the *ahorro* it generated.[49] Fidel Castro estimated that the country could "save" up to two-thirds of the oil metabolizing the socialist state with an energy system designed "rationally" to suit the post-Soviet geopolitical order. Such *ahorro* equaled 1.5 billion dollars on reduced oil imports, which in the mid-2000s could pay all state salaries twice over.[50] Energy conservation therefore surpassed oil as the country's most important energy resource, even if it was a resource that gave declining returns to investment: "We are not waiting for manna to fall from heaven and for lots of oil to appear," Fidel told the workers in UNE, "because we have discovered, fortunately, something much more important; the conservation of energy [*el ahorro de energía*], which is like finding a great oil reserve."[51]

By comparing energy conservation to a fossil fuel deposit, the government framed energy efficiency as a resource that could be tapped for economic growth. In contrast to an oil reserve, however, "saved" energy was an immaterial good, an imagined absence of energy-not-used that was produced by reconfiguring the infrastructural environment and changing human practices. As Sarah Knuth argues in relation to retrofitting programs in the United States, when it is framed as a resource, energy efficiency is imagined to be a "thing" that is "intangible but real, quantifiable, and potentially valuable."[52] The nominalized abstraction of something not expended can subsequently be managed for the purpose of reducing an energy deficit and enabling capital accumulation. Hence, in Cuba just as in the United States, energy conservation was not only imagined as a response to an imposed condition of scarcity but also as an opportunity for economic growth.[53]

To invest in energy efficiency, two international agreements provided capital for the Energy Revolution. First, Cuba's membership in PetroCaribe meant that the socialist state did not have to buy oil on the international markets since it arrived on "fair" terms from Venezuela. In the years leading up to the 2008 financial crisis, this left the state with a surplus of hard currency previously spent on oil, which could sponsor the Energy Revolution. Second, Cuba's relations with China had improved significantly since the end of the Cold War. In 1960, Cuba was the first Latin American country to recognize the People's Republic of China, and China soon became a key rice exporter to Cuba. With the Sino-Soviet split, however, the relationship deteriorated, notably as Cuba valued Soviet oil over Chinese rice. In 1966, China cut its rice export quota, and in the ensuing diplomatic conflict Fidel Castro denounced Chairman Mao as a "senile idiot."[54] These tensions persisted until the late 1980s when Cuba's relationship with the Soviet Union soured and China was isolated after the crackdown on Tiananmen Square. A series of Sino-Cuban state visits then took place before China extended an interest-free credit to Cuba for the sale of domestic appliances in the early 2000s. As a commercial opportunity with geopolitical implications for China, offering a long-term strategy for Cuban import substitution, a joint venture was formed for the assembly of household electronics in Havana.[55] By 2010, China had become Cuba's second-largest trading partner, following Venezuela.

The second reason behind the call for *ahorro* related to the technical overhaul of the SEN. The low availability factor after the "Antonio

Guiteras" breakdown meant that UNE could only supply a limited amount
of electricity at any one time. Energy conservation and efficiency gains
then reduced the demand for energy, evening out the discrepancy between
the available capacity and the load on the system, just as the scheduled
blackouts had done in the special period. In one typical account, UNE's
national inspector for energy demand explained in *Granma*, "We must
. . . maintain the efforts of *ahorro* in our workplaces and homes in order
to contribute to the stability of this vital service."[56] State publications also
gave a third reason for *ahorro*, namely the environmental benefits of energy conservation. To "save" oil was tantamount to decarbonizing energy
use and infrastructure. A brochure published by UNE to inform the public
of the infrastructural works stated that *ahorro* was vital in order to reduce
the ecological footprint of consumption. Framed as such, *ahorro* allowed
the country to reduce its contributions to climate change, environmental
pollution, and the depletion of global oil reserves.[57]

Finally, a fourth reason tied together the economic, technological, and
environmental dimensions in moral terms. For the Revolution to succeed
economically and environmentally, it was the moral responsibility of the
Cuban people to be thrifty and use energy "rationally." The newspaper *Juventud Rebelde* reported that "Cuba, like many other countries, has the
obligation to redesign its policies for economic and social development,
as well as the responsibility to foment a rational use of oil on which [that
development] is based."[58] While the PCC did not abandon the economic
growth imperative, it was a moral necessity for it to reevaluate its long-
held ideas about the links between energy use and social development.
From the national level, the moral responsibility of *ahorro* translated
down to smaller scales, becoming an ethic for every socialist citizen to
live up to. In Pinar del Río, *Guerrillero* noted, "We have the great responsibility and moral commitment to the *Comandante en Jefe* [Fidel Castro]
to turn Pinar del Río into a model of efficient energy use in the year of the
Energy Revolution in Cuba."[59]

The moral implications of the *ahorro* discourse were again tied back
to an idealist socialist rhetoric. The Energy Revolution followed a period
in Cuban public life that had seen a revival of the mass politics of the
Revolutionary Offensive and the Rectification process. In 1999, the Elián
González affair created a wave of popular mobilization on the island when

people turned out in large numbers seeking the return of a seven-year-old boy from custody in Florida. Capitalizing on the moment, Fidel Castro launched the "Battle of Ideas," a national campaign centering on a series of educational initiatives and seeking to revive the guevarista ideals of social duty and human development as equally if not more important than economic development.[60] The "battle" stood between the selfless revolutionary ideals of the New Man and the individualism of a corrupt capitalist consumer culture. Following the most difficult years of the special period, the Cuban leadership and many regular Cubans experienced the country to be going through a moral crisis.[61] The thieving of resources and the corruption documented in the previous chapters exemplified such widespread "immoral" behavior. When Cubans privately "resolved" their daily needs by illicit means but publicly defended revolutionary values, it reflected a prevalent *doble moral* (double standard).[62] With the Battle of Ideas, Fidel Castro sought to revive the high moral values of revolutionary life and reinforce what Antoni Kapcia has called the "moral impetus" of the Revolution.[63] Thus, via the *ahorro* discourse, the Energy Revolution was a direct cultural continuation of the Battle of Ideas.

At a moment of social mobilization, the *ahorro* discourse drew on the revolutionary conception of history. In *Visions of Power in Cuba*, Lillian Guerra shows how the revolutionary narrative logic developed in the 1960s to require all citizens to adopt a morality of "sacrifice" for the Revolution to succeed. It also demanded that the Cuban people be continuously mobilized in the revolutionary process since every revolutionary citizen's active "struggle" and "resistance" against the counterrevolution comprised the moral impetus of the Revolution.[64] This conflation of morality and temporality took human form in the myths surrounding Che Guevara and José Martí who were both portrayed to have fought selflessly for the Revolution, and it was abstracted in Guevara's vision of the New Man. With the insertion of *ahorro* into this narrative, the acts of energy conservation and work efficiency became morally charged revolutionary virtues providing momentum to history. As the scale-dependent subject of Cuban national history, the Cuban people had to work *efficiently* and struggle *selflessly* for the Revolution until all reactionary forces had been overcome. The government's calls for *ahorro* therefore aimed to reinforce the Revolution's moral impetus, reinvigorating the revolutionary process.

The political effectiveness of the *ahorro* concept rested in the way it articulated a concrete thermodynamic process, a physical process in which energy efficiency increases, as a thing with moral, economic, technological, and environmental meanings. Imagined as a resource, *ahorro* could be measured in exact terms of gigawatt-hours of renewable energy output, megawatts of installed distributed generators, grams of fuel input per kilowatt-hour output, and percent of reduced heat losses in transmission. These measures of energy efficiency translated into moral and political accomplishments, attesting to the government's ability to make historical progress through the act of *ahorro*. Here, energy efficiency worked in the same way as the measurements of the SEN's capacity (MW), length (km), and connectivity (percent) did during its construction, coproducing the infrastructural state as an object of knowledge. Thus, when Fidel Castro reported that the energy intensity of the state economy had decreased by more than 15 percent in the first quarter of 2006, this piece of thermodynamic knowledge translated into economic, technological, environmental, and moral progress.[65]

The geographer Andrew Barry argues that the manner in which energy conversions are measured is "an explicitly governmental and political matter."[66] Drawing on the work of Isabelle Stengers, he shows how the notion of energy itself is inseparable from the act of measurement. To assess the efficiency of an energy conversion, physicists must imagine an ideal system in which a perfect conversion takes place—a so-called Carnot cycle—which then allows them to compare the conversions in a real system with this ideal. Since the Carnot cycle is a "fictional device" conditioned by the second law of thermodynamics, the measurement of energy conversions "creates the object it measures."[67] Similarly, there is an epistemological need to define the boundaries that delimit an energy technology in order to measure its social and environmental impact. Consequently, it is "only by measuring an object within selected system boundaries that the object comes into being."[68] The meaning of energy, therefore, is conditioned by the situation in which it is measured, and Barry sees it as necessary to interrogate how the measurement of energy efficiency is subject to regulatory practices, monitoring, and organizational management as an object of government.

For Cara New Daggett, a moral code, conceptually similar but reflecting a different historical experience than the Cuban, lies at the heart of thermodynamic knowledge. She notes how many of the physicists who developed the concept of energy during the Industrial Revolution were devout Scottish Presbyterians. The formulation of the laws of thermodynamics, through their work on the steam engine, came to reflect key aspects of their faith. The first law of thermodynamics establishes that energy is a constant, like God, but as soon as it is put to work, as the second law states, its usefulness diminishes for the flawed humans trying to command it. While the Carnot cycle reflects a divine ideal, only real entropy-increasing thermodynamic cycles exist in the imperfect human world. Taking the argument to its full conclusion, Daggett argues that the measurement of energy efficiency thus translated into an ethics of sin and sloth, mandating that waste (entropy) be minimized by frugal life and hard efficient work.[69]

In the context of British industrialization, the thermodynamic knowledge associated with energy efficiency had a power-serving effect. When white Europeans, in light of a superior work ethic, were deemed to have reached a higher level of civilization than their colonial subjects, it legitimized the expansion of the British Empire as a civilizing mission as a law of nature. "Energy," Daggett writes, "became a traveling metaphor that reinforced the material and capitalist relations of empire."[70] Her conclusion is that a radical politics for a socially just low-carbon future therefore must be based on concepts other than energy, work, waste, and efficiency, seeing that these reflect a raced, imperial worldview inextricably woven into the globalized fossil fuel economy. The measurement of energy efficiency, in other words, is an object of government that cannot be dislodged from fossil fuels as a reserve of work potential. In Cuba, however, the quest for efficient energy use was embroiled not in a raced cosmology of civilizational progress but in a postcolonial imaginary. The Cuban government's obsession with energy efficiency reflected an eschatological understanding of history, on a par with the eschatology of Scottish Presbyterianism, but this was one in which thrift and efficient work management pointed toward human liberation in the Global South. The spectacle of the Energy Revolution, as a political performance, allowed the Cuban government to demonstrate that it lived up to a national revolutionary ethos on a historical mission to thus restore a sense of political efficacy in the country.

AHORRO, MONITORING, AND FLUORESCENT LIGHT

In the pursuit of *ahorro*, the government set out to transform energy practices in the state economy. In Pinar del Río, the university and local enterprises worked jointly to improve the efficiency of production. Using a model developed with inspiration from a Japanese management system, the university recruited postgraduate students from the provincial workforce to its MSc program in Energy Efficiency. In their dissertations, the students were assigned to implement the *Tecnología de Gestión Total Eficiente de Energía* (Technology for Total Efficient Energy Management) in the state company in which they were normally employed and to reflect on the process. The assignment itself was relatively straightforward: first, they prepared an inventory of all the energy sources used in the company, such as diesel, fuel oil, and electricity. They then ranked these in order of importance and added them up until 85 percent of all energy expenses were accounted for. The resulting list was deemed to represent the company's "key posts" (*puestos claves*). Second, the students convened an energy council (*consejo energético*), bringing together selected managers, shift leaders, technicians, Party members, union representatives, and any members of the mass organizations Brigadas Técnicas Juveniles (BTJ— Technical Youth Brigades) and Asociación Nacional de Innovadores y Racionalizadores (ANIR—National Association of Innovators and Rationalizers) in the company. The energy council would then come up with a plan for how the key posts could be used more efficiently. Since the start of the MSc in 2005, the process was, at least in theory, repeated every two to five years in each company.[71]

The MSc curriculum reiterated that *ahorro* should be considered the country's principal energy resource. "Efficiency has become the principal source of economic growth under the current circumstances," a dissertation developed in the Pinar del Río branch of UNE reported.[72] "*Ahorro* and sustained efficiency must constitute the main energy source [for development]," claimed another from the "Harlem" sugar mill.[73] The students learned that there were two courses of action to take for a company to increase its energy efficiency. On the one hand, technical innovations and retrofits had the greatest potential for long-term *ahorro* but required large financial investments. On the other, organizational changes had a lower cost but were more difficult to implement. The main problem of state socialist practice, one lecturer noted, had been the lack of conscience of energy efficiency and the failure to account for it in the planning process. Despondent, he added that the lack of morals was Cuba's primary challenge in the pursuit of *ahorro*: "The organizational problems are killing us. What do you do when it's the company director's car that is the number one key-post consumer?"

As the sole distributor of electricity, UNE imposed stricter monitoring in state companies, and the newspapers regularly reported on quota violations. In 2008, for example, *Granma* recorded that there had been thirteen acts of electricity overconsumption in Pinar del Río and sixteen in 2009.[74] The Directorate of Rational Energy Use within the provincial branch of UNE noted after "nightly inspections" that electric lights often were kept on overnight, and that some factories, such as Pinar del Río's brewery, violated restrictions on electricity use during peak hours.[75] To penalize these workplaces, UNE cut the power supply to them to compensate for the wasted kilowatt-hours. In December 2007, administrative areas in Hotel Pinar del Río were blacked out for three days, and the same measure was taken in workplaces in Bahía Honda, San Cristóbal, and Minas de Matahambre.[76] However, *Granma* reported that workers in the affected companies resisted the measures, which complicated matters for the utility: "When you decide to shut off one of four floors due to indiscipline, there are new deficiencies appearing on the other [floors]."[77]

Despite these concerns in the university and the utility, organizational change received only marginal attention in state publications. The Energy Revolution was presented as a process of technological change and innovation. When *Bohemia* listed the "Principal actions of the Energy Revolution in 2007," it emphasized the reconfiguration of the SEN, the replacement of thousands of water pumps in aqueducts and multistory buildings, and the installation of electric ovens in bakeries.[78] The technical upgrades fell into two categories. The first were upgrades that increased the systemic efficiency of energy use. In the tourism industry, the government installed 3,200 solar heaters, which would heat water for hotel guests—previously requiring electricity—by utilizing solar energy directly (Figure 16). Solar heaters were also fitted in nurseries, hospitals, and central kitchens to reduce the demand for electricity. Thousands of wind turbines were in turn

Figure 16. Solar heaters at Hotel Cayo Levisa, Pinar del Río. Photo by Gustav Cederlöf.

installed in the agricultural sector to mechanically power water pumps, displacing the use of electricity.[79] Again, these technologies increased the efficiency of energy use by redesigning the infrastructural systems altogether, shortening the chains of energy conversions.

The second category comprised upgrades that increased the economic efficiency of energy use. Fidel Castro emphasized changes in the transport industry. In his accountant-like rhetorical style, he reported on the delivery of new tractors, concrete mixing trucks, semitrailers for containers, and replacement motors for lorries in exact numbers. The interprovincial bus company Astro received a new fleet of blue-colored buses from China, and the *camellos* ("camels") that had served as public transport in Havana since the special period were replaced with a new bus fleet.[80] In a widely publicized event, UNE's linesmen received 101 new pickup trucks. Symbolically handing over the keys to the workers, Fidel detailed that the utility at that point disposed of 523 Soviet ZiL-130 trucks capable of running 1.6 km on one liter of gasoline (km/L). It had 309 jeeps (6 km/L),

346 GAZ trucks (1.8 km/L), 149 Kamaz trucks (1.8 km/L), and 139 MAZ trucks (c. 1.8 km/L). The new pickups from the Chinese Great Wall Motor Company, together with a host of other vehicles being imported at the time, would replace this oversized, outdated machinery. When Fidel Castro announced that the pickups could run 14 km on one liter of diesel, the audience broke into applause in recognition of such great *ahorro*.[81]

In households, the government replaced a large number of electrical appliances. Fidel Castro argued that Cuba was in a particularly good position to "save" energy in this manner compared to other countries. During a mass rally in Pinar del Río, he reported that there were 2.4 million antiquated refrigerators in Cuba's households and that they consumed up to five times the electricity of modern fridges. In Pinar del Río alone, there were 143,000 refrigerators, and 136,000 of these were of Soviet or prerevolutionary capitalist brands such as Inpud, Minsk, and Frigidaire. Together they consumed around 20 percent of all electricity generated in the province during peak hours.[82] *Guerrillero* later reported that 143,587 refrigerators had been replaced with new ones in Pinar del Río during 2006.[83] Alongside the refrigerators, the government distributed electric fans, air conditioners, TVs, water pumps, thermostats, and gaskets for pressure cookers and coffee makers in quantities ranging from the hundreds of thousands to the millions.[84] The socialist economy was becoming energy efficient. Most products were of Chinese origin as the government made use of the interest-free credit offered by China. During a state visit in 2003, Fidel Castro had inspected the assembly line of the TV-manufacturer Panda in Nanjing, and reportedly, one million ATEC-Panda TV sets were assembled by the joint venture Empresa de la Industria Electrónica (Company of the Electronics Industry) in Havana in time for the Energy Revolution.[85]

The moral implications of the replacements were made clear at an event in the Great Hall of the University of Havana. Halfway through his speech to the university's students, Fidel turned to a member of the audience, asking how many incandescent light bulbs she had at home. The student, who lived in rural Santa María, replied that her family had two lights of 60 W each, and she estimated that they kept one lit for twelve hours and the other for four hours a day. "Twelve and four, 16 hours; times 60, gives 960 watts," Fidel calculated, calling on his aide Enrique to bring the student two fluorescent "energy saving" light bulbs of 7 W each. The two fluorescent lights would consume 112 W each over the sixteen hours of use—well below the consumption of the incandescent light bulbs—and Fidel revealed that the government would be distributing two fluorescent bulbs to every household in the country, phasing out incandescent lights. Thus, the student's two lights were becoming fifteen million. Having handed her the bulbs, Fidel again turned to the student to thank her for *her* gift to the country. Through her energy use, interacting with the infrastructures of the socialist state, she would now contribute to *ahorro* and, hence, the success of the Revolution.[86]

To celebrate what it saw as the momentous success of the Energy Revolution, the government framed the campaign as a key event in the history of national liberation and socialist development. The Energy Revolution constituted a new epochal shift in the temporality of revolution. While there was an *antes* the Revolution (before 1959) denoting an epoch of underdevelopment, and an *antes* the special period (before 1990) defining a period of development, Fidel Castro announced that there would now be "an *antes* and a *después* [after] the energy revolution in Cuba."[87] There was no official announcement that the special period had ended, but the Energy Revolution most definitely marked its end. Later, the Central Bank of Cuba emblazoned the back of the 10 CUC banknote with an image representing the Energy Revolution so that the campaign was placed alongside the Battle of Santa Clara (where Che Guevara led the rebel army to victory) on the 3 CUC note and the Protest of Baraguá (where Antonio Maceo affirmed his determination to fight the Spanish crown) on the 5 CUC note to designate a milestone in the history of Cuban revolutionary struggle. In my conversation with the electrician Norberto, it was also clear that things no longer were like they had been: "So it's not like *antes* [before]?" I asked. "Ha-ha, *¡muchacho!* [boy]," he said. "No. It's not like *antes.*" After Fidel Castro had inaugurated the emplacement where we met, there were no longer any blackouts. We were meeting in a time after the special period; a time demarcated by the historical rupture of the Energy Revolution. "There haven't been any problems since they installed the generators," Magdalena said. "You know, things are much better. Well, it's still tough, but it's nothing like what it was *antes*. I remember when Fidel came here . . . now there aren't any *apagones* any longer."

ENERGY TRANSITION AS POLITICAL CENTRALIZATION

While replacing domestic appliances, the government also offered all households a new set of equipment: an electric pressure cooker, a rice cooker, an electric hotplate, a kettle, and an immersion heater (Figure 17). This was the first time new domestic appliances had been available after they were withdrawn from the state shops in 1990. Not for the first time, the household was a key space for the reconfiguration of the country's energy systems, and like before, the intervention had gendered implications. Politically, the distribution of electrical cookers was quite a U-turn—during the Soviet period, Fidel Castro had repeatedly stressed that electricity was unsuitable for cooking given Cuba's underdeveloped electrical industry. But now, electrical cooking appliances would enable Cuba's women to cook under modern, energy-efficient circumstances. In public appearances, Fidel Castro spoke of the pressure cooker as *La Reina* (the queen), and when I did fieldwork in Cuba even fifteen years after the Energy Revolution, these "new" appliances not only provided the infrastructure for cooking, but the pressure cooker was still affectionately referred to as *La Reina* in everyday conversations.[88]

The electrical appliances enacted an energy transition in households. The pressure and rice cookers permitted Cuban women, predominantly, to cook without kerosene, LPG, and denatured alcohol, or even the diesel, charcoal, wood, and sawdust that had been common fuels during the special period. They also relieved household members from the daily "struggle" of procuring these fuels. A man interviewed in *Bohemia* testified to this cause for celebration, stating that the electrical appliances "freed me of tremendous struggle. I spent all my time looking for fuels wherever I could. The *Pike* in my home put up with everything from crude oil to gasoline. My wife drove me crazy with the oil leaks and the soot."[89] Foremost, however, the appliances freed women from the arduous task of acquiring cooking fuels, as this primarily had been their responsibility.[90] In contrast to most announcements during the special period, state publications anticipated a move toward more modern life and increased productivity, suggesting that the new cooking appliances would save the population both time and energy. *Guerrillero* wrote that by automating the cooking process with inbuilt timers, the appliances made cooking simpler for women doing housework. Indeed, *La Reina* increased "the quality of family life."[91] Not

Figure 17. Electric rice cooker and *La Reina* pressure cooker. Photo by Gustav Cederlöf.

least important from the government's point of view, electricity use would reduce the population's exposure to noxious fumes, especially from diesel bought on the informal market. "Such are the effects of this energy revolution," Fidel Castro observed referring to the health outcomes, "of this saving [*ahorro*] of energy that we are talking about."[92]

However, just as Marisa Wilson finds in her study of everyday moral economies, some of my interlocutors were clearly discontent with their appliances.[93] Ivelis, for example, said that the cookers indeed were good to use, but to fry an egg or to make coffee, gas would have been more practical for her. She argued that gas also would have been better for the environment, since the hotplate was far too large for her *cafetera* (coffee maker). Angelo, who was a mechanical engineer, was upset about the equipment, as it was inefficient. The pots used with *La Reina* were made from aluminum and iron, and this did not suit the cathode in the cooker, he argued. It would have been better for the government to introduce equipment providing heat through induction instead, or to have switched to gas, which was both an efficient heat source and cheap to produce on the island. "But the decisions are made over our heads," he said tapping his shoulder with two fingers. "They [the government] don't listen either to the experts or the population." Thus, had it been optional, people may have chosen a different techno-energetic setting in their kitchens, but only if the technology was actually available and cost effective to them. The socialist state, not the market, allocated resources to the population.

By framing energy efficiency as a resource for investment and economic growth, a new frontier for resource appropriation was opened up inside homes and other privately owned spaces.[94] The government expanded its infrastructural power by appealing to efficiency gains and making households more dependent on electricity. At the same time, it intensified state control over the circulation of liquid fuels in the country. Framed as an initiative to clamp down on corruption, Fidel Castro dispatched thousands of *trabajadores sociales* (social workers) to monitor the distribution and commercialization of oil products. The *trabajadores sociales* were students who had been trained as part of the Battle of Ideas to form a kind of youth vanguard working to reinforce the Revolution's moral impetus. The students seized control over the country's gas stations and rode along in tankers hauling diesel and gasoline from ports and refineries. State media soon reported on extensive networks of black-market filling stations being discovered along the major roads.[95] It was of course well-known that such informal trade took place before the surveillance started and "discovered" here implies that these practices became legible for the state. After one month, state revenue from diesel and gasoline sales had reportedly increased 280 percent.[96]

The *trabajadores sociales* also revealed widespread theft of kerosene and diesel. Diesel had become a common cooking fuel during the special period when workers stole it from their workplaces and traded it on the informal market. Operating at odds with the de-commodifying, redistributive logic of the socialist state, many of these informal spheres of exchange were closed down through the increased monitoring of energy flows. Meanwhile, the government took kerosene, LPG, and denatured alcohol off the *libreta*, ceasing to distribute these liquid fuels in the rationing system. Prior to the Energy Revolution, each household in Pinar del Río had been entitled to 10 kg of LPG every eighteen days for the symbolic sum of 2.50 CUP, or an equivalent amount of kerosene or denatured alcohol. Seeing that Pinar del Río had been hit by several hurricanes in recent years, each household was still entitled to a limited supply of LPG as a safety precaution, but this amounted to no more than one 10 kg canister every six months for 10 pesos. Thereafter, it was possible to purchase extra canisters for more than 100 pesos a piece.[97] This was an astronomical sum for those employed in the state sector, such as one of my colleagues doing administrative work at the university who earned just over 300 pesos a month.[98] Thus, the state almost entirely uncoupled the population from one energy system based on liquid fossil fuels while incorporating them technologically in another based on electricity.

With their new electrical appliances, Cubans relied less on informal "immoral" networks supplying liquid fuels. However, the withdrawal of cooking fuels from the rationing system meant that it was almost impossible to refrain from electricity use in kitchens due to processes acting on a larger-than-household scale. Liquid fuels flowed in ever smaller quantities through the country and under more intense state monitoring. Between 2003 and 2009, official figures indicate that the gross household consumption of LPG, kerosene, and denatured alcohol decreased by 48, 55, and 69 percent, respectively, while the consumption of electricity went up by 25 percent.[99] Thus, while the government enabled electrified cooking, it infrastructurally foreclosed the possibility of cooking with energy sources other than electricity. At once, it intensified state-household interactions in the quite mundane act of cooking. In his study of British industrialization, Timothy Mitchell argues that coal workers had been able to make political gains by disrupting the supply of coal into the economy at infrastructural chokepoints, such as coalfaces and railroads. However, with the transition from coal to oil, these infrastructural spaces lost significance in view of the materially different oil infrastructure, foreclosing organized workers' action.[100] In Cuba, the transition to electricity similarly foreclosed informal energy trade, allowing the state to reestablish hegemony over resource distribution and combat selfish, iniquitous behavior. The electrification of cooking tied together and centrally coordinated energy use in the socialist state, materially and narratively reinforcing the political economy of its infrastructural form.

ENERGY VULNERABILITY AND AN "UPDATED" SOCIALIST STATE

The Energy Revolution opened up a new infrastructural space for the negotiation of state-citizen relations. To some in Pinar del Río, the newfound dependence on the SEN was a source of concern, because with experiences of hurricane damage fresh in mind, this arrangement increased their vulnerability to power cuts. Angelo explained that they had long relied on LPG for cooking in his neighborhood: "In the area where I live, we had liquefied gas for a long time. But now we get only one canister every six months. One ten-kilogram canister for 10 pesos. *Antes* [i.e., before the Energy Revolution], we received one canister every eighteen days and it cost 2.50. You can also buy extra canisters, but they cost over 100 pesos each." When asked, he confirmed that the single gas canister was meant as a safety precaution, only to be used under exceptional circumstances, but he argued that the small amount of gas they received therefore left them vulnerable. The dependence on electricity and the restricted access to gas left him with little autonomous control over his energy use: "It is an enormous problem. With the electricity infrastructure we have and the high dependence on electricity, there can be a power cut at any point and people can't prepare their food. What do you do then, without electricity and without gas?"

Angelo's worries also concerned the economic impact of energy use, and this concern was rooted in a more far-reaching reform process in the socialist state. From the outset of the Revolution, electricity had been construed as an entitlement to the socialist citizen. As chapter 2 showed, the socialist citizen received tokens in the form of Cuban pesos to acknowledge their labor contributions to the national economy, and by returning some of these to the state, they gained access to electricity as a non-commodified use value through state infrastructure. However, the special period made this principle of socialist energy justice difficult to maintain. First, Fidel Castro noted that the state now distributed electricity as a non-commodity without consideration of who the end-user was—whether a worker motivated by social duty or a corrupt petty capitalist: "The electricity is very cheap, it is given away; whether we give it to pensioners or workers, it is given away . . . but we are also giving it to the racketeer [*el merolico*] who charges 1,000 pesos from here [Havana] to Guantánamo, or charges twice the monthly salary of a doctor to bring him from Havana to Las Tunas, with robbed fuel."[101] Thus, the state had come to reciprocate not only moral but also immoral work with access to electricity. Second, the state distributed electricity as a use value to the "newly rich"—those running businesses in the legal private sector, making profits off wealthy foreigners. This meant that the state was subsidizing an emerging petty bourgeoisie. Finally, the state provided electricity to citizens who were fortunate to have friends or family overseas, remitting money to them. Remittances, Fidel argued, were not received in exchange for labor time in the Cuban economy, and therefore, the socialist state had come to subsidize electricity use unreciprocated by labor contributions to national production.[102]

To confront these issues, the government introduced a new electricity tariff in 2005. A progressive rate would serve three purposes: first, the state would stop subsidizing immoral electricity use; second, it would incentivize *ahorro*; and third, it would serve as a mechanism for income redistribution. The political leadership reasoned that the main electricity consumers were likely to be citizens working on *cuenta propia*—people running *casas particulares* (private houses) or *paladares* (private restaurants) and thereby keeping high-consuming kitchen appliances, electric showers, and air conditioners in their homes. A progressive tariff would then serve as a form of taxation so that people with a high income contributed more to the exchequer. The first 100 kWh were priced at 9 ¢/kWh before the rate increased in ten price bands from 30 ¢/kWh for consumption falling between 101 and 150 kWh up to 5 CUP/kWh for any consumption exceeding 5,000 kWh.[103] This contrasted with the tariff set in 1980 to a largely symbolic 6.5 ¢/kWh irrespective of consumption. With the new tariff, electricity was no longer construed only as a use value but was assigned exchange value proportional to purchasing power. Income, then, regulated the citizen's ability to access state infrastructure. This was a qualitatively different approach to the principle of socialist distribution based on the ideal of entitlements.

The new tariff was aimed at people like Anabel and Reinier, who ran *casas particulares*, renting out rooms to tourists as their main source of income. To gain a license for a *casa particular*, the government required homeowners to keep a TV, an air conditioner, an electric fan, and a refrigerator in every room for rent. As Table 1 shows, Anabel had several electrical appliances in her house, and in addition to these, her house guests brought iPads, digital cameras, and cell phones. One of Anabel's electricity bills from 2013 indicates that her average monthly consumption was 788.3 kWh or, converted into pesos, 1,035.60 CUP. Normally, she would charge 20–30 CUC per night for one of her two rooms, corresponding to 480–720 CUP, and this soon covered her high electricity costs.

Table 1   Electrical appliances in Anabel's house

Item	Quantity-Brand-Origin
Refrigerator	1x "Kelvinator" (US, 1950s)
	2x "Haier" (Energy Revolution, 2000s)
Air conditioner	1x Soviet (1980s)
	2x "LG" (Energy Revolution, 2000s)
TV	1x "Phillips" flat screen (2010s), bought during a trip abroad
	3x "ATEC-Panda" (Energy Revolution, 2000s)
Electric fan	4x "Rayfan," "Midea," "Daytron," unknown
La Reina pressure cooker	1x "Wanjiafu" (Energy Revolution, 2000s)
Electric rice cooker	1x "CHC-Nova" (Energy Revolution, 2000s)
Freezer	1x Soviet (1980s)
Microwave oven	1x "Sanyo"
Blender	1x Unknown
Washing machine	1x Unknown
Electric iron	1x "La Plancha" (Soviet, 1970s)
Hairdryer	1x "Conair" (2010s)
DVD player	1x "Phillips"
Radio clock	1x "Sony"
Cordless phone	2x "vtech"
Number presenter	1x "Red Lion" (broken)
Laptop	1x "Dell" (2010s)
Cell phone	2x Unknown
Fluorescent light bulbs	18x "Phillips," "LG"
Fluorescent tubes	14x "Phillips"

For state employees, in comparison, the new tariff made electricity the largest monthly household expense. One day, Ivelis asked me how much I paid for the room I was renting in Pinar del Río. Her monthly salary was 300 CUP, and converted into CUC, I paid almost twice this sum for my room per day. If my landlord had no problem paying his utility bill with this income, Ivelis's electricity consumption amounted to one-third of her monthly salary. "You know, my electricity bill is about 100 pesos a month," she said. "Fine, I have quite a few electrical appliances at home—but still." Her cooking appliances were the main sources of energy consumption, and LPG use under the economic conditions of old looked like a more attractive alternative: "Imagine the price difference compared to gas," she said.

On another occasion, I talked to Frida about potable water. I had just purchased a 5 L water container for a little under 2 CUC. This was well above what Frida could afford, testifying to my privileged position in the household. She explained that she rarely drank water but that her daughters put iodine in the tap water to kill bacteria. To boil it was out of the question. For this, they would have needed the hotplate, and Frida could not afford the electricity. Angelo was visibly upset when thinking about the electricity tariff. The government claimed that the tariff had been introduced so that people working on *cuenta propia* would pay more, he said. "But I don't work on *cuenta propia* and now I have to pay 570 pesos a month for my electricity use." He continued:

> You can't use electrical appliances just to enjoy yourself. But everyone has to cook. . . . In the evening, when it is hot up in the twelve-story building where I live, you should be able to turn the fan on for a while. If I can't sit in front of the fan for a while and drink some cold water, how am I supposed to get up in the morning and manage to work? It's the electricity bill that drains the household budget. I pay around 570 pesos, and fair enough, I have several electrical appliances at home, but that's as much as the doctors have been paid until recently!

The new tariff aimed to even out socioeconomic differences emerging in the aftermath of the special period, but the state-citizen relations negotiated at the point of energy use instead appeared to reinforce inequalities. Paired with the electrification of cooking, the new tariff skewed income differences. As Stefan Bouzarovski and colleagues point out in a study of energy vulnerability in Central and Eastern Europe, structural shifts in energy systems are linked to processes of economic and social differentiation so that energy transitions can "render some actors more socially and economically vulnerable . . . creating new inequalities across time and space."[104] In Cuba, the Energy Revolution altered the political economy governing energy use, which differentiated the population's ability to engage in everyday energy use. Those on a low income, often deriving from state employment, had to spend a large share of it on electricity, while *cuentapropistas*—people with a private-sector income—could increase their energy consumption with hard-currency earnings in Cuba's dual economy. Thus, the levels of energy vulnerability increased among a part of the population.

As an infrastructural intervention changing the way national and international processes shaped local experiences of energy use, the Energy Revolution had profound outcomes. Materially, it decarbonized the Cuban economy in relative terms, reducing its carbon intensity with one-third and almost halving its energy intensity. While it stabilized the throughput of energy in the infrastructural state, the Energy Revolution also made the SEN—and hence Cuban socioeconomic life—more resilient to the impact of hurricanes. It did so by closing down many of the alternative, locally constituted infrastructural systems that had enabled energy use during the special period. In the midst of such infrastructural transformation, the Energy Revolution also reconfigured core political economic principles. When the government granted citizens access to energy relative to income, the state ceased to be the provider of energy as an entitlement and instead became a service provider, distributing electricity based on purchasing power. Thus, the Energy Revolution paved the way for some of the most radical reforms initiated by Raúl Castro, intending to "make structural changes and [changes] of concept" in the socialist state.[105] In the aftermath of the Energy Revolution, the socialist state is less the provider of social wealth and more the provider of services to an individualized body politic for whom work efficiency is identified as a revolutionary imperative.

Conclusion

ENERGY TRANSITIONS AND INFRASTRUCTURAL FORM

Late in the afternoon, the lights went out in Havana's Vedado district. "*¡Ay, caramba!*" Anabel cursed in front of the television. Two decades after the power cut at the beginning of this book, the source of the disruption could still not be located among the fuses—this, Anabel asked me to check. The day was hot and so was the drinking water. Later in the evening, we sat with two of Anabel's friends under the starlit Havana sky, enjoying the cool air and sweet *tilo* tea.

"I wonder why there was an *apagón*," Esme said.

"They never tell us anything," Diancy snapped, referring to the government.

"Maybe it was something with the oil again?" Esme asked.

Wise in the affairs of her street, Anabel let us know that it was not the oil this time. She had spoken to the UNE linesmen earlier that day, and they had been cutting down branches over the distribution cables to pre-empt a blackout.

"They cut the power in the whole area."

In the years following the Energy Revolution, Cubans saw a period of relative stability in their energy supplies. For the government under Raúl Castro, efficient consumption and secure energy supplies were integral to a program of "updating" and "perfecting" Cuba's socialist system. In 2014, the PCC declared in unmistakable style that "efficient energy generation constitutes one of the driving forces behind the structural transformations that are being carried out through the implementation of the Guidelines approved by the Sixth Congress of the Party."[1] When hurricane Irma left large parts of Cuba in ruins in 2017, it toppled utility poles and pylons in the central provinces and seriously damaged the pump house at the "Antonio Guiteras" thermoelectric plant. But Cuba's distributed electricity system was still able to supply emergency power, a colleague relayed from Pinar del Río: "At the moment the entire province is supplied with electricity from the diesel generators."

However, the death of Hugo Chávez and the presidency of Donald Trump again changed the geopolitical landscape in the Caribbean and, with it, the energy situation.[2] The collapse of the international oil markets in 2015 exacerbated an already downward trend in the Venezuelan oil industry, and President Trump's sanctions on PDVSA aggravated the situation. Venezuela's plummeting oil output and political instability has affected Cuba and the PetroCaribe alliance dearly. The Cuban government's most recent figures indicate that the country's crude oil imports fell by a whopping 45 percent between 2015 and 2019.[3] In September 2019, Raúl's successor President Miguel Díaz-Canel headed a meeting in Pinar del Río reporting that 89 percent of the restaurants and canteens in the province had resorted to cooking with wood or charcoal. Animal traction also played an increasing role in transports of people, goods, and waste. Rolling blackouts were again a factor in everyday life.

In the new situation, *ahorro* remained a key ideological motif. Once Cubans had seen through the crisis, Díaz-Canel argued in Pinar del Río, it would be important to retain some of the alternative solutions that had come back to use from the most difficult years of the special period: "We can't go back to being inefficient," he said. "We have to work with . . . consistency so that the fuel that arrives in the future lasts us longer."[4] Nevertheless, a state commission led by the minister of energy and mines and a senior member of the politburo concluded that fuel theft remained one of the country's principal problems. The tension between revolutionary morale and illicit everyday practices endured.[5] For an outside observer, it is tempting to compare the situation with the onset of the special period, claiming that history repeats itself. Yet the Cuban economy has been far less dependent on Venezuelan oil than it ever was on the Soviet counterpart, and Cuba's uneven low-carbon transition has reduced the country's overall dependence on oil as well as its vulnerability to geopolitical shocks.[6] Even so, the energy situation today is acute, and despite the system of distributed generation, the SEN suffered a nationwide blackout during the calamitous hurricane Ian in 2022.

In 2014, the PCC adopted a new energy policy. With 95 percent of the country's electricity still generated from fossil fuels, the Party announced a renewed transformation of the SEN. Based on a 3.5-billion-dollar investment, renewables are planned to make up 24 percent of the SEN's total capacity in 2030. The policy is a revitalized effort to decarbonize the grid but, above all, to establish an energy system territorialized on the Cuban islands. The low-carbon transition reduces international dependencies and increases national sovereignty. Thus far, three trends stand out. First, UNE is focusing almost exclusively on solar and wind power. In 2019, *Juventud Rebelde* reported that sixty-five solar parks had been synchronized with the SEN and that fifteen more were under construction.[7] With Chinese financing, UNE is also expanding the existing wind parks in Gibara and Isla de la Juventud. In an effort to adapt the SEN to generation from renewables, designed as it is for thermoelectric and nuclear production, a European company has advised the Ministry of Energy and Mines (MINEM) on options for energy storage under the auspices of the Paris Agreement. A consultant with the firm, whom I met in London, explained that MINEM only supplied the bare minimum of information and that its key priority was to secure hardware, software, and know-how so that Cuban engineers later could work independently of international contacts.

Second, AZCuba, the successor institution of the Ministry of Sugar (MINAZ), seeks to increase electricity production from biomass. While bagasse remains an important fuel, it is the invasive species *marabú* (*Dichrostachys cinerea*) that attracts the most attention. Native to sub-Saharan Africa, the Spanish introduced this thorny bush to the Caribbean in the sixteenth century after which it quickly colonized the Cuban countryside. With access to synthetic herbicides from the Soviet Union, MINAZ and the Ministry of Agriculture prevented *marabú* from spreading with some success, but during the special period, and especially once MINAZ laid vast expanses of cane fields fallow after 2002, *marabú* colonized the countryside without restraint.[8] AZCuba has now partnered with the British-Chinese utility Havana Energy to develop harvesting machinery and a first bioelectric plant located in Ciego de Ávila. A director of the new plant, operated by the joint venture Biopower S.A., explained that three tons of *marabú* could generate electricity equaling one ton of fuel oil, and the plant could therefore "save" Cuba up to 100,000 bpd. The government plans to build twenty-six bioelectric plants of the kind, but according to the plant director, this would only be possible with "a lot more foreign investment."

Finally, the 2030 policy is enhancing the commodification of energy embarked on with the Energy Revolution. Havana Energy and the French utility Hive are both building utility-scale solar parks in Cuba's recently opened special development zone in Mariel. Exempt from normal Cuban legislation in the economic enclave, the utilities are able to retain full ownership over infrastructure and to sell electricity into the surrounding socialist economy for a profit. In 2019, Decree-Law 345 also established that Cubans can purchase photovoltaic solar panels for private use and sell energy surpluses to UNE. This integrates households into the infrastructural state in a new way, now as producers rather than consumers of a good traded with the state utility. In a brief on the new regulations, MINEM noted that the law would reduce carbon emissions from diesel use in the electricity sector, but more importantly, it would undermine the United States' sanctions on PDVSA and PetroCaribe. Solar energy, the ministry argued, would be a weapon against "the new pirates of the Caribbean ... who at any cost chase the oil tankers bound for Cuban soil." Still, the cost of photovoltaic panels would make autonomous electricity generation an option mainly for those working in the private sector, thus further differentiating energy access on the basis of socioeconomic standing.[9] Energy use, and the prospect of a low-carbon energy transition, therefore remains a key area in which the political, economic, and symbolic circumstances of life in Cuba are negotiated.

IMAGINARIES OF DEVELOPMENT, JUSTICE, AND ENERGY TRANSITION

One of the aims of this book has been to examine how an imaginary of energy use and postcolonial development has been placed in Cuban politics, and to explore how this imaginary has been renegotiated in periods of geopolitical and environmental change. Scholars working outside Cuba have often taken the collapse of the Soviet Union as an analytical starting point, interpreting Cuba's recent history variously through a lens of sustainable development or successful degrowth.[10] The reduced energy and carbon intensity of the national economy; the decentralization of the energy systems; the efforts of energy conservation (*ahorro*); and the widespread turn to direct energy applications undoubtedly offer many insights on possible degrowth trajectories. With a longer-term perspective on Cuban history, however, it is evident that Cuba's enforced low-carbon situation was narrated differently inside the country. Inserting Cuban experiences in a degrowth narrative risks doing epistemic violence to these experiences. A revised ecological concept of Cuban state socialism has instead developed that remains firmly rooted in a political project in which growth is considered a sine qua non for historical progress.

From the 1960s, energy use, and especially electricity consumption, played a leading role in Cuba's revolutionary drama as it unfolded toward an overdetermined future. Centralized infrastructures allowed the socialist state to redistribute energy for the purpose of economic development and social equalization. The infrastructural state itself emerged as a vehicle of socio-ecological transformation, operationalizing a notion of energy justice: centralized infrastructure was an enabling condition for national development and socialism.[11] Prior to the Revolution, Cuba's electricity and gas infrastructures primarily served affluent urban communities and the sugar industry. But the revolutionary government envisioned a unified, territorially encompassing energy infrastructure to serve the interests of an unclassed, deracialized, and non-gendered community—the collective-singular Cuban people. Thus, the SEN and the rationing system concretized a scalar strategy, providing infrastructural form for a national political project.[12]

When electricity, oil, and nuclear energy were placed at the center of the socialist project, an imaginary of growth-based development also put key human-environment relations at stake. Marx identifies labor as "a condition for human existence which is independent of all forms of society," and he notes that humans employ machines and energy extracted from the environment to transform nature through labor.[13] The socialist economy was predicated on a transition from organic energy use—energy harnessed in fields and forests—to the use of vertically extracted energy resources. An oil-fueled and ultimately nuclear economy would allow mechanical work to replace manual labor. The transition to communism was also an energy transition: the aim was to transform Cuba from an economy associated with human toil, underdevelopment, and colonial subjugation into an electric economy, freeing the independent Cuban people from work. Socialism hinged on automated power replacing muscular power—machines liberating men and, with time, women doing housework.

The experience of energy system change cannot be understood without reference to the influence of gendered social relations. In the 1950s, private investments in electrification were seen to be profitable only if they were met by increasing energy demand. The uneven geographical development of the electricity sector therefore reflected the utilities' ability to turn a profit.[14] As chapter 1 showed, households were a key space in which the utilities could create demand by appealing to a gendered housewife ideal. After the Revolution, the socialist state instead allocated resources through centralized planning, undermining the law of value, and in what followed, the government prioritized industrial applications over investments in stoves, kettles, and other domestic appliances. The configuration of the techno-energetic environment in households was the basis for experiences differentiated by gender in the special period. Cuban men and women did not share the same spaces for infrastructural interactions, and consequently, they had different experiences of the frequent blackouts and the shortages of cooking fuels. Moreover, the Energy Revolution's success, not only as a technological but as a cultural intervention, can be explained in relation to how the state again intervened in gendered household relations. The government articulated the electrification of cooking as a process in which the Revolution modernized household practices, increased energy efficiency, and thus improved the quality of life for Cuban women.

To understand how an imaginary of energy use and development was renegotiated during the special period, it is also necessary to go beyond the household level and examine how the oil shortages called for new ways of thinking about socialism as a national project. As chapters 4 and 5 demonstrated, the SEN remained a fossil-fuel-dependent energy system, and efforts of low-carbon transition largely occurred alongside the centralized infrastructures, through bricolage, informal trade, and regulated market exchange. As Iris Borowy argues, Cubans were compelled "to live according to degrowth rules" in the special period. They had to "produce and consume locally, refrain from credits, change from energy-intensive mechanized to low-energy, labor-intensive production methods, replace long distance with face-to-face communication and live a simple, low consumption life-style."[15] In a radical shift from the modernist socialism of the 1980s, the economy deindustrialized in the 1990s. Encouraged by the PCC, many state companies rid themselves of machinery dependent on centralized fuel distribution. De-mechanization went counter to the prevailing understanding of socialist development, but despite this, it did not undermine the revolutionary master narrative. Instead, deindustrialization was folded into the history of revolution through the "special period" as a spatiotemporal category of experience. The special period was a politically necessary moment of exception in a longer trajectory of development and liberation.

The survival of the socialist state has depended on the malleability of the revolutionary narrative. When the economy contracted and people sought alternative energy solutions, the government promoted systemically more efficient, nonstate infrastructures. However, they only did so if energy remained a non-commodity, and they only tolerated a nonstate solution if it left the state's distributional monopoly over the surplus product intact. The continued legitimacy of the state also depended on the translatability of the revolutionary narrative across scales. The vernacular of "innovation," "thrift," and "resolving things" resonated with a national history of "struggle" and "resistance" so that everyday practices were negotiated with a master narrative of postcolonial socialist development: through the act of energy use, the socialist citizen did something for the Revolution and the success of the state. Popular accounts of the Cuban Revolution often articulate a strict binary between "the state" and "the people," but this binary must be seen to have emerged historically as an effect of human-infrastructure interactions in Cuba.[16] In the most general sense, energy system change was productive of citizen and state as evolving subject positions.

If Cuba's post-Soviet history is interpreted in a degrowth frame, this warrants an important question: what subject positions do degrowth practices call for? Writing on degrowth and labor organization, Stefania Barca argues that degrowth needs "a clearer vision of what political subjects and which processes of political subjectivation can make it happen."[17] The subject positions emerging from the state-socialist paradigm were closely tied to a particular vision of energy use. Oil was a steppingstone to a nuclear energy system, and the dialectic between the productive forces' development and Cuba's social development centered on a human subject liberated from work.[18] In this process, the state played a coordinating role in deciding how the products of labor should be managed, and in the Marxist-Leninist tradition, the state had to reinvest surpluses in order to enable the transition to communism.

The alternative energy solutions that appeared in the special period materially de-grew the Cuban economy, making its metabolism smaller, but they did not generate new subject positions with socially transformative implications. Instead, they were articulated either as extraordinary achievements by a collective-singular revolutionary people who resisted the imperialist reaction or as individually corrupt and immoral behavior. The alternative practices thereby folded into a morally charged narrative centered on the state and its citizens in which the "special period" was but a hiatus in a longer trajectory of socialist development. In this narrative, gendered and racialized everyday experiences were dissolved in a collective ethos of struggle. A core challenge for degrowth, then, is how people negotiate their own place in a new political economy, but most importantly, how new political subjects can emerge through praxis. While Cuba's special period was a low-carbon situation produced by geopolitical upheaval, degrowth must emerge through voluntary practices of self-limitation that generate new political subjectivities.

CONTESTED ENERGY GEOGRAPHIES: INFRASTRUCTURE, CENTRALIZATION, AND AUTONOMY

A second aim of the book has been to explore how the uses of particular energy sources, such as oil and electricity, give rise to spatial patterns of social and economic activity that engender political ideals, social relations, and economic interests. A key question for research on energy transitions is how low-carbon development upsets the spatial patterns fossil fuels have enabled and requires their reconfiguration, often in socially unequal and politically contentious ways.[19] In the Cuban context, it is difficult to dissociate this question from the process of state formation and the role of energy infrastructures in carrying state projects. Here, I have taken energy use as a starting point to understand it as a culturally situated socio-ecological practice in relation to events and processes that take place on a far larger scale than the practice itself.[20] The situated practice must then be conceptualized together with social relations of production and repro-duction, state formation processes, and geopolitical changes as a conti-nuity rather than a series of discreetly spatialized phenomena, which an energy transition reconfigures in its totality.[21]

If we follow Beatriz's work with her Aurika washing machine (chap-ter 3) through the socket and into the grid so that "plugs and sockets come to mark connections along a circuit, not impenetrable interfaces between inside and outside," the situated practice is enabled by an infrastructure that exceeds the space of human experience.[22] This infrastructure itself embodies complex narratives, geopolitical and biophysical relations that act on larger scales and connect back to the social practices they enable. Energy use, in other words, has infrastructural form. The state-socialist aim was to distribute energy evenly as an entitlement to a territorially dispersed citizenry, but with energy use as the starting point, the state cannot be understood as a reified outcome: in Cuba, it was less an auton-omous agent holding infrastructural power than an effect of infrastructure and culturally negotiated energy use.[23] From this perspective, the infra-structural power of the state is best understood as the ability to maintain a territorialized political-economic arrangement of resource distribution, which enables human action and interaction.[24] Infrastructural mainte-nance, like that undertaken during the Energy Revolution, is an effort to maintain such a political economic arrangement.[25]

Moving further along the circuit, it is necessary to situate the state in the international political economy in order to explain its territorial ex-pansion and fragmentation. As an infrastructural phenomenon, grid-based electrification spatially separates the practice of energy use from the environment sustaining it, displacing the environmental load of de-velopment to elsewhere. The centralizing logic of the socialist state thus depended on a continuous supply of low-entropy energy and raw materials from overseas. Thus, if technological systems rely on a net social appropriation of resources, as I have argued with Alf Hornborg and Andreas Roos, "the image of a self-sustained machine is as misleading as it would be to imagine an organism surviving without continuous inputs of nutrients."[26] Historically, a key question for the Cuban government has been what ability the national community has to determine the conditions upon which energy is supplied to the island. When the government nationalized the oil refineries in 1960, it exhibited its power over foreign interests to control key nodes in the energy system. When it partnered first with the Soviet Union and later with Venezuela, it showcased how Cuba was able to enter into foreign exchange on what it deemed to be non-exploitative terms. The collapse of the Cold War geopolitical order, however, fundamentally upset the conditions upon which Cuba was inserted in the international political economy, which undermined the state's centralizing, redistributive logic.

In the special period, Cubans were challenged not only with finding innovative solutions to immediate problems but also with reconfiguring long-established spatial patterns of social and economic activity. When the infrastructural state failed, manifold economic activities were re-embedded in local ecologies. The new infrastructures were often place specific, and they generated geographically uneven development in terms of energy availability and energy access. For example, chapter 4 showed how charcoal was commodified and traded locally in Pinar del Río, creating new social relations that undermined the state's resource monopoly. The energy crisis thus reconfigured the established geographies of energy distribution, which challenged basic tenets of Cuban socialism. Highly developed, spatially concentrated and interconnected productive forces, as it were, enabled the state to redistribute value rising from human, ecological, and technological action at scale. But in the low-carbon special period, immediate application and local autonomy were more highly prioritized political goals than redistribution and interconnectivity over a large area.

A political geographical dynamic pitting redistribution against autonomy has pulled the infrastructural state in contradictory directions in the post-Soviet period. On the one hand, the government has acted on the basis of territorial integration, economies of scale, and a centralization of control. These factors have been preconditions for economic growth and a decommodification of energy. On the other, a countercurrent of territorial fragmentation, a deconcentration of infrastructural capacity, and a decentralization of control has run through the state. The special period's alternative energy systems increased territorial self-sufficiency and allowed Cubans to withdraw from state interactions. However, these opposing forces have been far from clear cut. The deconcentration of generation with diesel and fuel-oil generators during the Energy Revolution increased municipal autonomy but, at the same time, recentralized control over energy use in the state. Through the coincident electrification of Cuba's kitchens, the Energy Revolution made households more dependent on extra-local infrastructure, thus increasing their vulnerability to energy disruption.

While energy transitions fundamentally rework spatial relations, the geographical forms of energy technologies do not determine political outcomes. Instead, it is necessary to evaluate how certain energy sources and technologies are mobilized as part of wider projects with political outcomes in mind.[27] In Cuba, the countercurrents continue to create waves in a debate on the country's future direction. Here, the large-scale, Sovietized projects of the past are giving way to a development strategy increasingly focused on self-sufficiency and autonomy. In one interview, an energy expert in Pinar del Río explained that "historically, there has been a mentality of centralization, but this is slowly changing. And this is empowering the region and cooperative work." The key issue today, she argued, was to reduce import dependencies and increase local autonomy. Another centrally placed energy scholar argued that "we are moving towards much cleaner energy use; we are moving towards more independent energy use. . . . The main thing is; it is that a country first of all has to be independent."[28] Nothing less than the implications of energy justice are at stake in these discussions: Does energy justice entail equality in distribution or local autonomy? Is energy justice tantamount to centralization or decentralization, even geographical development or territorial self-determination? These are all progressive political goals in a context shaped by colonization, and implicitly or explicitly, they must be negotiated during an energy transition.[29]

MORE THAN (DE)GROWTH: ENERGY SOVEREIGNTY
AND POSTCOLONIAL ENERGY FUTURES

While Cuban discussions today center on different geographical priorities in the process of low-carbon development, broader debates on energy transitions tend to focus on another contested matter: the need for economic growth or degrowth. Arguments that emphasize the importance of growth come in neoliberal, Keynesian, and socialist guises, and they all indicate that growth is necessary for bringing about qualitative social and technological changes at scale.[30] Growth, it is assumed, is the source of abundance. In Cuba, as we have seen, there is a long history of deploying energy-intensive technological solutions to overcome the limits of local ecologies for growth and socialist development. By contrast, degrowth calls for a radical reorganization of socioeconomic life so that the economy becomes more decentralized and is regulated by self-imposed limits. The degrowth hypothesis holds that economic growth destroys an already abundant planet, and by collectively imposing limits on growth, communities create new opportunities for enjoying that abundance.[31] Paradoxically, Cuban history also showcases initiatives that attest to this ideal: energy development attuned to geopolitically imposed limits.

Anglo-American scholars arguing for the imperative of growth have recently made a case for solar energy as a material condition for socialist development. A Marxist interpretation of infrastructure sees it as a store of past productive potential—"a means for gathering and holding productive powers in technological suspension"—and as such, it enables qualitatively new processes to unfold.[32] Fossil-fuel–based production can therefore possibly create the conditions of possibility for a technologically more complex, sustainable solar future. In a polemic with degrowth scholars, Matthew Huber fears a return to intensive agricultural labor as part of a degrowth alternative (a fear he presumably reserves for populations in the Global North), and he argues that fossil fuels can be a historical "bridge" to a sustainable yet energy-intensive socialist alternative.[33] The argument for such a techno-utopian solution to climate change crystallizes most clearly in calls for "solar communism." Solar communism, David Schwarzman argues, denotes a high-energy economy metabolized by "soon-to-be-developed technologies" in which solar energy enables the increased throughput of materials in the "technosphere" without adverse effects on the biosphere.[34] Crucially, reflecting a dialectical conception of history, solar communism does not represent a retreat to a preindustrial organic economy, but "a revolutionary leap forward into a new form of sociality, one that is energy intensive and technologically enabled."[35]

When ecomodernist socialist theories like these are put forward, they resonate with the debates that have unfolded in Cuba and the socialist world. The vision of a society that overcomes environmental limits through technological action justified the construction of the SEN in the 1960s, Cuba's nuclear program in the 1980s, and the Energy Revolution in the 2000s. It also guided more controversial environmental projects not covered here, such as the revolutionary plans to drain the Zapata Swamp, to generate artificial rain, and to breed a Cuban super cow—what Reinaldo Funes Monzote identifies as an ambition of *geotransformación*.[36] However, to overcome biophysical limits and sustain the socialist economy with low-entropy supplies, Cuba has needed to engage in uneven geopolitical relations, extending its energy systems beyond the absolute territorial boundaries of the island. Indeed, as an anti-colonial project, the Cuban Revolution was itself an attempt to seize popular control and renegotiate the terms on which Cuba was incorporated in the neocolonial world economy; an economy in which other populations had drawn on Cuban resources to overcome the limits of their immediate environments.[37]

By contrast, the inability to access fossil energy in the post-Soviet period triggered both national and local initiatives to territorialize an energy regime on the Cuban islands. Self-imposing limits on the economy's geographical extent, these initiatives were attempts to adapt to, rather than overcome, biophysical limitations, and they illustrate how the Cuban vision of modernist socialism had been contingent on the ability to displace the environmental load of development. They therefore open up important questions for ecomodernist projects today: for whom is eco-socialist development possible? Who can undertake a Keynesian-inspired Green New Deal? Indeed, to embark on a progressive political project based on growth, how must the community doing so be placed in the global political economy for it to be possible? Any attempt at low-carbon transition must be attentive to the social relations that govern the unequal distribution of resources and risk in the world economy.[38] Thus, if we start theorizing energy transitions from a Caribbean instead of an often-unacknowledged Euro-American experience, energy transitions must be seen as processes that intervene in and reconfigure colonially patterned geographies.[39] No account of energy transition can avoid addressing how global and local power geometries shape their socio-ecological conditions, albeit from different historical-geographical vantage points.

Cuban history provides a useful outlook for social science energy research, not least as it throws the Western capitalist mainstream into sharp relief: Cuba's history is both socialist and anti-colonial. At a distance, it is easy to envision the Cuban islands as bounded spaces; as geophysical formations with a finite resource base, defined by their difference from places exterior to them. However, as Antonio Núñez Jiménez argued in *Geografía de Cuba*, it is a "simplistic idea" to suggest that "Cuba is an Island."[40] The development of the national economy was enabled by material and symbolic exchanges that extended beyond Cuban shores. As Conor Harrison and Jeff Popke write in their work on Caribbean energy geographies, islands are "a mix of closure and openness," and extra-islandic exchanges take place between actors with historically produced, unequal power resources.[41] Key questions for research on energy transitions, then, focus on how energy systems connect and disconnect places and populations: On what terms? On whose terms? What dependencies are created? And from an anti-colonial or even decolonial viewpoint, how are energy systems enrolled in politics of territorial sovereignty?

These questions are of immediate relevance in Cuba today where the energy policy adopted in 2014 takes a significant step toward a more sovereign low-carbon economy. In the special period, Cuba was geopolitically isolated and had to embark on a low-carbon transition. Today, it is pursuing an energy transition in order to isolate itself. However, the government does not strive for autarky, or self-sufficiency within a bounded space. Cuban energy sovereignty rather concerns the power to control the relations through which a place enters into wider systems of material and symbolic exchange. This is the ability to maintain a particular political economic arrangement of resource flows—infrastructural power in other words.[42] Conceptually, connectivity and isolation can be regarded as two sides of the same coin: if involuntary isolation is to be excluded from connection, voluntary isolation is the ability to exclude oneself from connection. Social power is maintained in the ability to isolate and to connect, and an energy transition is inherently about this mode of power.[43]

Even as Cuba invests in renewables, the PCC predicts that the consumption of fossil fuels will increase until 2030, thus shortcutting the transition in absolute terms. The growth imperative guiding the socialist economy—as indeed any capitalist economy—will create a rebound effect so that the share of renewables and the consumption of fossil fuels will increase simultaneously due to growing electricity output. Investments in renewables only contribute to a real low-carbon transition if fossil fuels are taken out of use at the same rate. Such a prospect, however, goes against the modernist growth-oriented paradigm long dominating Cuban socialism. Even so, Cubans have developed an extensive low-carbon economy since the special period, reworking geopolitical relations, not least in the agricultural sector where wind turbines, solar heaters, and agroecological methods are in widespread use.[44] Given these conflictual patterns, can we speak of a "Cuban miracle" as so many observers have? The answer depends on which factors we emphasize in a history that is not free from contradictions. While oil scarcity is the mirror image of oil dependence, a low-carbon economy is defined in relation to an economy locked into fossil fuel use. To envision a radical energy future, the challenge is not to negate what already exists, but to break out of the binary.

Notes

PREFACE

1. All names of people appearing in the book, other than those in official positions, are pseudonyms.

2. Bridge, "The Map Is Not the Territory," 16; see also Baka and Vaishnava, "The Evolving Borderland of Energy Geographies"; Newell, "Race and the Politics of Energy Transitions"; and Sovacool, "What Are We Doing Here?" for a review of the skewed geographies of social science energy research.

3. Lambe and Bustamante, "Cuba's Revolution from Within," 21.

4. Siebert et al., *Four Theories of the Press*, 5.

5. Sparks, *Communism, Capitalism and the Mass Media*.

6. Lambe and Bustamante, "Cuba's Revolution from Within," 11.

INTRODUCTION

1. *Guerrillero*, "Informan sobre reajustes en la programación de apagones."

2. Castro Ruz, *La historia me absolverá*, 53.

3. Castro Ruz, "Discurso pronunciado . . . en la concentración popular efectuada en la Plaza de la Revolución 'José Martí', en Honor del compañero Leonid Ilich Brezhnev, Secretario General del Comité Central del Partido Comunista de la Unión Soviética, y la delegación que lo acompaña."

4. Larkin, "The Politics and Poetics of Infrastructure."

5. ONE, *Estadísticas energéticas en la Revolución*, tables 14, 25; ONEI, *Anuario estadístico de Cuba 2014*, tables 10.14; 10.17.

6. See, e.g., Castro Ruz, "Discurso pronunciado . . . en el Acto Central por el XXX Aniversario de los Comités de Defensa de la Revolución, efectuado en el Teatro 'Carlos Marx.'"

7. Molina Pérez, "De luz y de sombra," 4.

8. Urry, *Societies beyond Oil*, 215.

9. WWF, *Living Planet Report*.

10. Hickel, "The Sustainable Development Index." In 2017, the World Bank revised its methodology for calculating the purchasing power of a Cuban income, and after a 2021 update of the SDI, Cuba ranked ninth globally; see Sustainable Development Index.org, "Methodology and Data."

11. Piercy et al., "Planning for Peak Oil," 170. The Cuban peak oil narrative is presented in a number of studies, including but not limited to Friedrichs, "Global Energy Crunch"; Levins, *Talking about Trees*; and Pfeiffer, *Eating Fossil Fuels*, as well as the noted documentary *The Power of Community: How Cuba Survived Peak Oil* (2006) directed by Faith Morgan.

12. Deere, "Cuba's National Food Program and Its Prospects for Food Security."

13. Hedges et al., "Epidemic Optic and Peripheral Neuropathy in Cuba."

14. Sánchez Serra, "La agricultura urbana, suburbana y familiar."

15. Murphy, *Cultivating Havana*, 9. On rural agricultural transformation, see Funes et al., *Sustainable Agriculture and Resistance*; Wright, *Sustainable Agriculture and Food Security in an Era of Oil Scarcity*; and Rosset et al., "The Campesino-to-Campesino Agroecology Movement of ANAP in Cuba." On urban farming and its institutionalization, see Altieri et al., "The Greening of the 'Barrios'"; Cederlöf, "Low-Carbon Food Supply"; and Premat, *Sowing Change*. Beside the focus on agroecology, there is a rich literature on the development of progressive environmental regulation in the 1990s, especially around the implementation of Agenda 21, see Bell, "Environmental Justice in Cuba"; Saney, *Cuba*; and Stricker, *Toward a Culture of Nature*, and on landscape and coastal conservation, see Gebelein, *A Geographic Perspective of Cuban Landscapes* and Whittle and Rey Santos, "Protecting Cuba's Environment."

16. Smil, *Growth*, 492.

17. Kallis et al., "Research on Degrowth," 296; see also, e.g., Hickel and Kallis, "Is Green Growth Possible?"

18. Gudynas, "Beyond Varieties of Development."

19. Hickel, "The Sustainable Development Index." Hickel's observation also finds support in other recent studies seeking to internalize the environmental impact of economic growth in indices of human development, see, e.g., O'Neill et al., "A Good Life for All within Planetary Boundaries" and Zhang and Zhu, "Incorporating 'Relative' Ecological Impacts into Human Development Evaluation."

20. Latouche, *Farewell to Growth*, 9; see also D'Alisa et al., *Degrowth*; Healy et al., "From Ecological Modernization to Socially Sustainable Economic Degrowth"; Hornborg, *Nature, Society, and Justice in the Anthropocene*; Kallis, "In Defence of Degrowth"; and Paulson, "Degrowth."

21. Demaria and Kothari, "The Post-Development Dictionary Agenda," 2594.

22. Meadows et al., *The Limits to Growth*.

23. Kallis, *Limits*.

24. For studies discussing degrowth in Cuba, see, e.g., Boillat et al., "What Economic Democracy for Degrowth?"; Borowy, "Degrowth and Public Health in Cuba"; Cederlöf, "Low-Carbon Food Supply"; Gerber, "Degrowth and Critical Agrarian Studies"; and Kallis et al., "Research on Degrowth."

25. Paulson, "Degrowth."

26. Knuth et al., "Rethinking Climate Futures through Urban Fabrics," 1336; Robbins, "Is Less More . . . or Is More Less?"

27. Key works on British industrialization include Malm, *Fossil Capital*; Sieferle, *The Subterranean Forest*; and Wrigley, *The Path to Sustained Growth*. On the United States: Black, *Petrolia*; Huber, *Lifeblood*; and Nye, *Electrifying America*.

28. See Miller and Warde, "Energy Transitions as Environmental Events," 467 for a critique of the historiography of energy transitions and Barnes and Floor, "Rural Energy in Developing Countries"; Crosby, *Children of the Sun*; and McNeill, "The First Hundred Thousand Years" for stadial interpretations.

29. Wrigley, *The Path to Sustained Growth*; Dukes, "Burning Buried Sunshine"; Sieferle, *The Subterranean Forest*.

30. Mitchell, *Carbon Democracy*, 15.

31. Nye, *Consuming Power*, 191.

32. While the figures are not entirely reliable, the forests were estimated to have decreased from 409,825 to 250,845 *caballerías* (1 *caballería* equals 13.42 ha). For a full discussion about the levels of deforestation and the uncertainties involved, see Funes Monzote, *From Rainforest to Cane Field in Cuba*, 131–33.

33. Huber and McCarthy, "Beyond the Subterranean Energy Regime?"

34. Smil, *Power Density*, 255; see also Bridge et al., "Geographies of Energy Transition."

35. Global coal consumption peaked at 162.58 EJ in 2014 and has remained relatively steady since then, reaching 160.10 EJ in 2021. This was 2.72 times the consumption in 1965 and 1.62 times the consumption in 2000; BP, *Statistical Review of World Energy* ("Coal: Consumption"). See Barak, *Powering Empire* for an analysis of the role of coal in the global economy.

36. Miller and Warde, "Energy Transitions as Environmental Events," 469; Edgerton, *The Shock of the Old*; McNeill and Engelke, *The Great Acceleration*. See also Huber and McCarthy, "Beyond the Subterranean Energy Regime?" and Malm, *Fossil Capital* for important critiques of Wrigley. From a less systemic perspective, Masera et al., "From Linear Fuel Switching to Multiple Cooking Strategies," 2083 refer to the parallel use of old and new energy sources as

"household accumulation of energy options," arguing against a stadial model of energy transition (see note 28).

37. Hornborg et al., "Has Cuba Exposed the Myth of 'Free' Solar Power?"

38. Castán Broto, *Urban Energy Landscapes*; Hornborg, *Nature, Society, and Justice in the Anthropocene*; Mitchell, *Carbon Democracy*; Pomeranz, *The Great Divergence*.

39. Georgescu-Roegen, *The Entropy Law and the Economic Process*; Kondepudi and Prigogine, *Modern Thermodynamics*. See Daggett, *The Birth of Energy* and Rabinbach, *The Human Motor* for a critical history of thermodynamics.

40. Carse, *Beyond the Big Ditch*, 11.

41. Adams, *Energy and Structure*; Bunker, *Underdeveloping the Amazon*; Hornborg, *The Power of the Machine*. See also Fischhendler et al., "Light at the End of the Panel"; Overland, "The Geopolitics of Renewable Energy"; and Vakulchuk et al., "Renewable Energy and Geopolitics" on nascent research on the geopolitics of renewable energy.

42. Cederlöf and Kingsbury, "On PetroCaribe"; Cederlöf, "Out of Steam"; see also Massey, *For Space*.

43. Anand et al., *The Promise of Infrastructure*; Larkin, "The Politics and Poetics of Infrastructure"; Star, "The Ethnography of Infrastructure."

44. Larkin, "The Politics and Poetics of Infrastructure," 328.

45. Clark, "Island Development," 609; see also Cederlöf and Hornborg, "System Boundaries as Epistemological and Ethnographic Problems" and Malm, "No Island Is an Island." For anthropologists, Malinowski's study of the Kula ring, a system of gift exchange among the inhabitants of the Trobriand Islands, is the paradigmatic case. Harrison and Popke, "Geographies of Renewable Energy Transition in the Caribbean" capture the absolute and relational geographical dynamics of island life in terms of the "island energy metabolism."

46. Massey, "Power-Geometry and a Progressive Sense of Place."

47. Hornborg, *The Power of the Machine*; Hornborg, *Global Magic*; Hornborg, *Nature, Society, and Justice in the Anthropocene*. See Cabrera Arús, "The Material Promise of Socialist Modernity" and Hernández-Reguant, "Inventor, Machine, and New Man" on material culture and modernity in Cuba.

48. Hornborg, *Nature, Society, and Justice in the Anthropocene*, 147.

49. Ingold, "From Science to Art and Back Again."

50. Hornborg, *The Power of the Machine*, 1.

51. Mann, "The Autonomous Power of the State."

52. Kale, *Electrifying India*.

53. Power and Kirshner, "Powering the State."

54. The notion of the autonomous state is particularly strong in work that approaches energy transitions from the perspective of international political economy, see Newell, *"Trasformismo* or Transformation?" The view of the state as an environmental actor was also important to the development of political ecology. Notably, Blaikie and Brookfield, *Land Degradation and Society* argued

that local experiences of environmental change should be interpreted through a "chain of explanation" in which the "local" is explained in relation to state regulation and knowledge, and further up the chain, a state's position in the global political economy. See Bryant and Bailey, *Third World Political Ecology* and Robbins, "The State in Political Ecology" for a fuller account.

55. Mitchell, "Society, Economy, and the State Effect"; Meehan, "Tool Power"; Painter, "Rethinking Territory"; Painter, "Prosaic Geographies of Stateness"; Power and Kirshner, "Powering the State."

56. See Cederlöf, "Maintaining Power"; Harris, "State as Socionatural Effect"; and Parenti, "The Environment Making State" on the state as a socio-ecological relation, and von Schnitzler, "Traveling Technologies" and Meehan, "Tool-Power" on infrastructure as a terrain for the contestation of state power.

57. Robbins, "Is Less More . . . or Is More Less?"; see also Boyer, "Infrastructure, Potential Energy, Revolution" and Huber, "Fossilized Liberation."

58. Gómez-Baggethun, "More Is More," 3; Knuth et al., "Rethinking Climate Futures through Urban Fabrics," 1338.

59. Coopersmith, *The Electrification of Russia, 1880–1926*; Högselius, *Red Gas*; Schmid, *Producing Power*; see also Hughes, *Networks of Power*; Hecht, *The Radiance of France*; and Phalkey, *Atomic State*.

60. Castán Broto, *Urban Energy Landscapes*, 23; von Schnitzler, "Infrastructure, Apartheid Technopolitics, and Temporalities of 'Transition'," 135; see also Boyer, "Energopower" and Hecht, "Introduction."

61. Larkin, "Promising Forms," 184.

62. Harvey, "Infrastructures in and out of Time," 84; see also Simone, "Infrastructure" and Gupta, "The Future in Ruins."

63. Boyer, "Energopower"; Huber, *Lifeblood*; Princen et al., *Ending the Fossil Fuel Era*.

64. See Heffron and McCauley, "The Concept of Energy Justice across the Disciplines"; Hornborg et al., "Has Cuba Exposed the Myth of 'Free' Solar Power?"; Jenkins et al., "Energy Justice"; and Sovacool and Dworkin, "Energy Justice" on the notion of energy justice.

65. An important exception is the now extensive literature on the political economy of energy transitions in southern Africa, see, e.g., Baker et al., "The Political Economy of Energy Transitions"; Castán Broto et al., "Energy Justice and Sustainability Transitions in Mozambique"; and Power and Kirshner, "Powering the State."

1. AGAINST THE ENERGY EMPIRE

1. Núñez Jiménez, *Geografía de Cuba*, 5, 222. See Funes Monzote, *"Geotransformación"* on the work of Núñez Jiménez and the controversy surrounding the first edition of *Geografía de Cuba*.

2. Gettig, "Oil and Revolution in Cuba."

3. Mitchell, *Carbon Democracy*, 39–40; see also Huber, *Lifeblood*, 44–57 on how scarcity was produced in the US oil industry in the 1930s in the face of oil glut, and Bridge and Wood, "Less Is More" for a more general examination of the socio-technical production of scarcity in relation to recent debates on peak oil.

4. Gettig, "Oil and Revolution in Cuba," 8; see also Cederlöf and Kingsbury, "On PetroCaribe."

5. Hurricanes are a key theme in a growing literature on Cuban environmental history. See especially Johnson, *Climate and Catastrophe in Cuba and the Atlantic World in the Age of Revolution* and Pérez, *Winds of Change*.

6. Moreno Fraginals, *The Sugarmill*; McGillivray, *Blazing Cane*.

7. Funes Monzote, *From Rainforest to Cane Field in Cuba*.

8. Moreno Fraginals, *The Sugarmill*, 76.

9. Crosby, *Ecological Imperialism*, 127–28.

10. Zanetti and García, *Sugar and Railroads*, 4.

11. On sugar and centralization, see Scott, *Seeing Like a State*, 394n17.

12. See Pérez, *On Becoming Cuban* for a rich study of US-Cuban cultural and political relations from the 1860s until the 1959 Revolution.

13. McGillivray, *Blazing Cane*; Yaffe, *Che Guevara*, 7, 172.

14. García, *Beyond the Walled City*.

15. Altshuler and González, *Una luz que llegó para quedarse*, 183–96; Anonymous, "Lo que fué y lo que es la . . . Cía, Cubana de Electricidad," 819. See Zanetti and García, *Sugar and Railroads* on the history of railroads and Altshuler, *Las comunicaciones internacionales de Cuba* on the history of telecommunications in Cuba.

16. Altshuler and González, *Una luz que llegó para quedarse*, 201–9. See Hughes, *Networks of Power* on Edison's work.

17. Altshuler and González, *Una luz que llegó para quedarse*, 301–4; Pérez, *On Becoming Cuban*.

18. Kapcia, *Cuba*.

19. Pérez, *Cuba*, 135–44. The Platt Amendment also gave the United States access to the area around Guantánamo Bay, which it has occupied ever since.

20. Hausman and Neufeld, "The Rise and Fall of The American and Foreign Power Company."

21. O'Brien, "The Revolutionary Mission," 765, 772.

22. Rodríguez Castellón, "La industria eléctrica cubana," 155. In the late 1950s, the Occidental Electricity System extended from Los Palacios in Pinar del Río to Nuevitas in Camagüey and the Oriental Electricity System from Guantánamo to Manzanillo in Oriente.

23. See further Walker, *Energy and Rhythm*.

24. Peters, "Electrification of the Hershey Cuban Railway"; Winpenny, "Milton S. Hershey Ventures into Cuban Sugar." See also Zanetti and García, *Sugar and Railroads*. "The Sweetest Place on Earth" is the registered trademark of Hershey, PA, see www.hersheypa.com.

25. Pérez, *On Becoming Cuban*, 329–30.

26. Harrison, "The Historical-Geographical Construction of Power," 479. See also Hirsh, *Power Loss* and Howell, "Powering 'Progress'" on the history of investments and utility regulation, and Nye, *Electrifying America* for a cultural history of electrification in the United States.

27. Pérez, *On Becoming Cuban*, 329–30.

28. IBRD, *Report on Cuba*, 325.

29. *Revolución*, "Su preferencia por este monograma está plenamente justificada . . . ," 16.

30. *Revolución*, "Lo mejor para mamá," 6.

31. *Revolución*, "Mayo 10: Día de las Madres," 23.

32. *Revolución*, "La Cía. Electric de Cuba se enorgullece en brindarle cocinas y calentadores cubanos Wesco," 7.

33. Pérez, *On Becoming Cuban*, 6.

34. Fulgencio Batista was a general and formerly elected president who came to power in a military coup in 1952 with the support of the United States.

35. Yaffe, *Che Guevara*, 9. Income elasticity denotes how much the demand for a product increases when the income of the consumers demanding the product increases.

36. Incidentally, Prebisch put forth the argument at an ECLA conference held in Havana in 1949; see Toye and Toye, "The Origins and Interpretation of the Prebisch-Singer Thesis."

37. Peet and Hartwick, *Theories of Development*, 64–68.

38. Núñez Jiménez, *Geografía de Cuba*, 24; see Galeano, *Open Veins of Latin America* for a classic exposition of the argument.

39. Alexander, *A History of Organized Labor in Cuba*; Pérez-Stable, *The Cuban Revolution*.

40. O'Brien, "The Revolutionary Mission." President Carlos Mendieta, who came to power with the support of Batista, returned control over the Compañía Cubana de Electricidad to AFP.

41. Pérez-Stable, *The Cuban Revolution*, 71.

42. Alexander, *A History of Organized Labor in Cuba*; Córdova, *Castro and the Cuban Labour Movement*. Notably, David Salvador was among those removed and imprisoned. In 1947, PSP supporters had in turn been purged from the CTC, see Pérez-Stable, *The Cuban Revolution*, 73.

43. *Revolución*, "Rebajan tarifas eléctricas," 1; *Revolución*, "Van a saber lo que es una Revolución."

44. Alexander, *A History of Organized Labor in Cuba*, 202.

45. Administración Revolucionaria, "A todos los trabajadores eléctricos," 11.

46. *Revolución*, "Cofiño, Fraginals y la Embajada yanqui"; *Revolución*, "Van a saber lo que es una Revolución."

47. Castro Ruz, "Discurso pronunciado . . . en la Asamblea General de los Trabajadores de Plantas Eléctricas."

48. *Revolución*, "Expulsa el 26-7 a Amaury Fraginals y a F. Iglesias," 1.

49. Pérez-Stable, *The Cuban Revolution*, 95; Pérez-López, *The Economics of Cuban Sugar*, 12–13.

50. Yaffe, *Che Guevara*, 164–65.

51. Guevara, "Informe del Dr. Ernesto Che Guevara, Ministro de Industrias," 118.

52. Prijodko and Marcer, "La industria electroenergética soviética," 60.

53. Marx, *The Poverty of Philosophy*, 49. The question of whether Karl Marx was a technological determinist has been one of endless debate, see Heilbroner, "Do Machines Make History?" for an influential determinist interpretation and Bimber, "Three Faces of Technological Determinism" for a critique. The debate has largely suffered from a fetishization of machines, failing to recognize the political economy of metabolic flows they embody (see Hornborg, *Global Magic*).

54. E.g., Lenin, "Eighth All-Russia Congress of Soviets, 29 December."

55. Prijodko and Marcer, "La industria electroenergética soviética," 60, original emphases. A *sovkhoz* was a state farm and a *kolkhoz* a collective farm in the Soviet Union.

56. Coopersmith, *The Electrification of Russia, 1880–1926*.

57. Stites, *Revolutionary Dreams*, 49.

58. Scott, "High Modernist Social Engineering," 10–14 traces Lenin's fascination with centralized planning and automated technology to the German Empire's industrial mobilization during the First World War. The industrial coordination of the German war effort testified to how monopoly capitalism revolutionized the productive forces. At the time of the Bolshevik Revolution, Lenin pictured how the same planning techniques could be applied to the new Soviet state, which would become one integrated, disciplined socio-technical whole serving the long-term interest of the proletariat.

59. Lenin, "Eighth All-Russia Congress of Soviets, 29 December," 516–17.

60. Banerjee, "Electricity," 50.

61. Coopersmith, *The Electrification of Russia, 1880–1926*.

62. Castro Ruz, "Total la mecanización de nuestra zafra en 2 años," 3.

63. Enviados especiales de Novosti, "Bratsk," 6.

64. Marx, *Capital*, 772–81. Harvey, *Limits to Capital* draws out the full theoretical implications of this argument, showing how investments and devaluation of fixed capital in different geographic regions become a "spatial fix" for capital to overcome periodic crises of overaccumulation.

65. Yaffe, *Che Guevara*, 194.

66. Guevara, "Antonio Guiteras," 622; see also Castro Ruz, "Total la mecanización de nuestra zafra en 2 años," 3.

67. Guevara, "Reuniones Bimestrales," 149 quoted in Yaffe, *Che Guevara*, 163.

68. Guevara, "Discurso en el Primer Fórum de Energía Eléctrica."

69. Wrigley, *The Path to Sustained Growth*; see also Huber and McCarthy, "Beyond the Subterranean Energy Regime?"

70. See Cederlöf, "The Revolutionary City," 62–63 and Hernández-Reguant, "Inventor, Machine, and New Man" for a full exposition of the argument in the Cuban context. The argument is rooted in Marx, *Grundrisse*, 704–6 and *Capital*, 553–64.

71. Pons Duarte, *Política energética, política económica y desarrollo*. In *Lifeblood*, Matthew Huber shows how gasoline has enabled the reproduction of neoliberal visions of freedom in the United States, allowing people to navigate a suburban landscape embodying ideals of self-fulfillment. In Cuba and the United States, petroleum entered into starkly different political economic relations wrapped up in visions of human liberation, providing form for human life.

72. Silverman, *Man and Socialism in Cuba*; Yaffe, *Che Guevara*.

73. Wendling, *Karl Marx on Technology and Alienation*.

74. Wendling, 168, 177. See Hornborg et al., "Has Cuba Exposed the Myth of 'Free' Solar Power?" for an argument against such a Cartesian perspective on technology.

75. Irizarry Mora, *Fuentes energéticas*. Smith-Nonini, "The Debt/Energy Nexus behind Puerto Rico's Long Blackout," 71 shows how Washington politicians also understood the electrification of Puerto Rico as "a capitalist showcase and counterpoint" to the electrification of Cuba. Both Josephson, *Industrialized Nature* and Scott, *Seeing Like a State* highlight more broadly that Soviet and American engineers often cooperated on projects throughout the Cold War period and kept a close eye on each other's scientific-technical breakthroughs to confirm that they were on the right development track.

76. Fitzpatrick, *The Russian Revolution*; Sutela, *Economic Thought and Economic Reform in the Soviet Union*.

77. Sutela; Schroeder, "Economic Reform of Socialism."

78. Yaffe, *Che Guevara*, 63–67, 249.

79. Yaffe, 54, 243–44.

80. Yaffe, 57–58.

81. Guerra, *Visions of Power in Cuba*, 209.

82. Eckstein, *Back from the Future*, 38–41.

83. For a detailed review of Cuban political institutions and their history, see Bengelsdorf, *The Problem of Democracy in Cuba* and Roman, *People's Power*.

84. Cederlöf and Kingsbury, "On PetroCaribe," 127–28.

85. *Revolución*, "La primera gran zancadilla contra nuestra Revolución," 20; see also *Revolución*, "Agresión política a Cuba la actitud del 'trust petrolero.'"

86. Shirley and Kammen, "Renewable Energy Sector Development in the Caribbean," 244.

87. Folch, "Surveillance and State Violence in Stroessner's Paraguay"; Klingensmith, *"One Valley and a Thousand"*; Mitchell, *Rule of Experts*.

88. Castro Ruz, "Discurso pronunciado ... en el acto celebrado con motivo de la terminación del montaje de una Unidad en Tallapiedra de la Empresa Eléctrica."

89. Semevskiy, "The Problem of Cuba's Energy Supplies," 25–26.

90. Guerra, *Visions of Power in Cuba*, 87–88.

91. *Revolución*, "24 millones ahorrará Cuba en compra de petróleo."

92. *Revolución*, "La primera gran zancadilla contra nuestra Revolución"; *Revolución*, "Cuba revolucionaria ganará también la batalla de los abastecimientos." Rupprecht, "Socialist High Modernity and Global Stagnation" gives an important account of the cross-curtain relations between Brazil and the Soviet Union. Rupprecht's work, together with Högselius, *Red Gas* in the European context, is important as it complicates conventional bipolar accounts of the Cold War.

93. Morley, *Imperial State and Revolution*, 105; see also Maurer, *The Empire Trap*, 314–27.

94. *Revolución*, "La primera gran zancadilla contra nuestra Revolución," 1.

95. *Revolución*, "Orden a la Texaco de procesar petróleo del Estado"; *Revolución*, "Orden de refinar a la Esso y la Shell"; *INRA*, "¡Se llamaban!"

96. Gustafson, *Crisis amid Plenty*, 63–64.

97. Högselius, *Red Gas*.

98. Klinghoffer, *The Soviet Union and International Oil Politics*, 228; Rupprecht, "Socialist High Modernity and Global Stagnation"; Maurer, *The Empire Trap*, 332–36. By the mid-1980s, India was nonetheless a major importer of Soviet oil, see Mehrotra, *India and the Soviet Union*.

99. Bogomólov, *La industria energética sin crisis*; Lavigne, *International Political Economy and Socialism*; Reisinger, *Energy and the Soviet Bloc*.

100. Mesa-Lago, "Assessing Economic and Social Performance in the Cuban Transition of the 1990s," 866.

101. Mesa-Lago, 868.

102. Ratnieks, "Baltic Oil Prospects and Problems," 315. See Högselius, *Red Gas* for a detailed account of Soviet hydrocarbon exports to Central and Western Europe during the Cold War, which is the source of many European countries' import dependence on Russian gas today.

103. Cederlöf, "Out of Steam."

2. ELECTRIFICATION OR DEATH

1. Guevara, "Conferencia en el ciclo 'Economía y Planificación.'"

2. Guevara, "Discurso en el Primer Fórum de Energía Eléctrica."

3. Yaffe, *We Are Cuba!*, 101, 298n39.

4. Gumá, "100 mil kilovatios más para el consumo del país."

5. de la Torre, "El Partido en la construcción de la termoeléctrica de Mariel."

6. Semevskiy, "The Problem of Cuba's Energy Supplies," 30; Camacho, "Nunca antes un técnico tuvo tantas oportunidades como ahora.—Domenech."

7. Camacho, "Nunca antes un técnico tuvo tantas oportunidades como ahora. —Domenech."

8. The copy of Wettstein's travel memoir I consulted had been a gift to Antonio Núñez Jiménez, and it is clear from the text that Wettstein greatly sympathized with the Revolution.

9. Cederlöf, "The Revolutionary City"; see also Barkin, "Confronting the Separation of Town and Country in Cuba" and Susman, "Spatial Equality in Cuba."

10. Wettstein, *Vivir en Revolución*, 187.

11. Castro Ruz, "Discurso pronunciado . . . de Inauguración del Segundo Bloque de 100 Megawatts de la Termoeléctrica 'Máximo Gómez', de Mariel, La Habana"; Castro Ruz, "Discurso pronunciado . . . en el acto de Inauguración de la Termoeléctrica 'Carlos Manuel de Céspedes', celebrado en Ocasión de Conmemorarse el Día del Constructor"; see also Rodríguez Castellón, "La industria eléctrica cubana," 155.

12. PCC, *I Congreso del PCC*, El desarrollo de la industria §26.

13. PCC, *II Congreso del Partido Comunista de Cuba (Informe central)*, 7.

14. Castro Ruz, "Discurso pronunciado . . . en el acto celebrado con motivo de la terminación del montaje de una Unidad en Tallapiedra de la Empresa Eléctrica."

15. Castro Ruz, *La historia me absolverá*, 53.

16. del Barrio Menéndez, "Unifican sistemas energéticos de occidente y oriente."

17. PCC, *II Congreso del Partido Comunista de Cuba (Informe central)*, 7. Heat losses in transmission are explained in greater detail later in the text.

18. Oramas, "Se extiende ya de punta a punta de Cuba el sistema electroenergético nacional."

19. Guevara, "Discurso en el Primer Fórum de Energía Eléctrica."

20. Guevara; Castro Ruz, "Discurso pronunciado . . . de Inauguración del Segundo Bloque de 100 Megawatts de la Termoeléctrica 'Máximo Gómez', de Mariel, La Habana."

21. In comparison, thirty-five kilometers north of the city of Pinar del Río, a 16 kW mini-hydroelectric operated on the San Vicente River when I visited it in 2009. The plant opened from a private investment in the late 1910s at which time it also comprised two diesel generators imported from the United States. Initially, the plant served a regional system that stretched along the Pinar del Río north coast to Bahía Honda, but when San Vicente was incorporated in the SEN after the Revolution, the diesel generators were deemed inefficient and were decommissioned, just as the "Eliseo Caamaño" was years later. Gaiga, *La cruz al pie de los mogotes*, 51; personal communications.

22. Castro Ruz, *La historia me absolverá*, 49.

23. Guerra, *Visions of Power in Cuba*, 50; see also Castro Ruz, *La historia me absolverá*, 176n262. Guerra does not specify the remaining 3.9 percent.

24. Guevara, "Informe del Dr. Ernesto Che Guevara, Ministro de Industrias," 117.

25. Castro Ruz, *La historia me absolverá*, 49–50.

26. Castro Ruz, 50.

27. Cederlöf, "The Revolutionary City," 58–59.

28. Alvarez, *Cuba's Agricultural Sector*, 41–44; MacEwan, *Revolution and Economic Development in Cuba*, 206–9.

29. Guevara, "Discurso en el Primer Fórum de Energía Eléctrica."

30. PCC, *II Congreso del Partido Comunista de Cuba (Informe central)*, 13.

31. Cabrera Arús, "The Material Promise of Socialist Modernity"; see also Rosendahl, *Inside the Revolution*, 37–38 on the *emulación socialista*.

32. Castro Ruz, "Discurso pronunciado . . . en el acto celebrado con motivo de la terminación del montaje de una Unidad en Tallapiedra de la Empresa Eléctrica."

33. ONE, *Estadísticas energéticas en la Revolución*, tables 41, 57. This is not to say that electricity was never used for cooking. For example, in Lewis et al., *Four Women*, 212—a classic ethnobiography from the Cuban 1970s—twenty-seven-year-old Gracia explains how her family pays a relatively high electricity bill, as they have an electric hotplate and a sewing machine in their home. Hotplates, however, would have been purchased before 1959.

34. Cederlöf, "The Revolutionary City," 58; Wilson, *Everyday Moral Economies*.

35. Rosendahl, *Inside the Revolution*, 30.

36. Wilson, *Everyday Moral Economies*, xxi.

37. Swyngedouw, *Liquid Power*, 30.

38. Huber, *Lifeblood*, 5, original emphasis.

39. Castro Ruz, "Discurso pronunciado . . . en el acto celebrado con motivo de la terminación del montaje de una Unidad en Tallapiedra de la Empresa Eléctrica."

40. Anusas and Ingold, "The Charge against Electricity," 542.

41. Dove and Kammen, *Science, Society and the Environment*, 86–87.

42. Harvey, "Infrastructures in and out of Time," 95.

43. Hornborg, "Undermining Modernity," 197–98.

44. Harvey, "Infrastructures in and out of Time," 94; see also Jasanoff, *States of Knowledge*.

45. Castro Ruz, "Discurso pronunciado . . . de Inauguración del Segundo Bloque de 100 Megawatts de la Termoeléctrica 'Máximo Gómez', de Mariel, La Habana." Almost any speech or news report referring to the electrical sector demonstrates the same phenomenon, e.g. Castro Ruz, "Discurso pronunciado . . . en el acto de Clausura del Primer Fórum Nacional de Energía."

46. Mann, "The Autonomous Power of the State," 190.

47. Hecht, "Introduction," 2; Hecht, *The Radiance of France*; Power and Kirshner, "Powering the State"; Harris, "State as Socionatural Effect."

48. Graham quoted in Schwenkel, "The Current Never Stops," 103.

49. Schwenkel, 103. In Albania, road construction enrolled the population in the construction of socialist development. While roads testified to an automotive communist future painted by First Secretary Enver Hoxha's government, it was a future conditioned by the scarcity of cars in Albania at the time, see Dalakoglou, *The Road*.

50. Boyer, "Energopower."

51. Oramas, "Se extiende ya de punta a punta de Cuba el sistema electroenergético nacional."

52. Oramas.

53. Rodríguez Castellón, "La industria eléctrica cubana," 159; Castro Ruz, "Discurso pronunciado . . . en el acto de Clausura del Primer Fórum Nacional de Energía."

54. Castro Ruz, "Discurso pronunciado . . . en el acto de Clausura del Primer Fórum Nacional de Energía."

55. Castro Ruz; see also Pons Duarte, *Política energética, política económica y desarrollo*, 91–92.

56. Castro Ruz, "Discurso pronunciado . . . en el acto central por el XXXI Aniversario del Asalto al Cuartel Moncada"; see also Castro Ruz, "Discurso pronunciado . . . en el acto de Clausura del Primer Fórum Nacional de Energía."

57. Castro Ruz, "Discurso pronunciado . . . en el acto celebrado con motivo de la terminación del montaje de una Unidad en Tallapiedra de la Empresa Eléctrica."

58. Cooper, "Petroleum Displacement in the Soviet Economy," 378.

59. Gustafson, *Crisis amid Plenty*, 29, 36.

60. Andropov quoted in Gustafson, 43.

61. Cooper, "Petroleum Displacement in the Soviet Economy"; Gustafson, 29.

62. Gustafson, *Crisis amid Plenty* provides the most comprehensive analysis of the Soviet oil crisis and the campaign for fuel switching and energy conservation. Notably, bottlenecks in gas storage and the Chernobyl meltdown hampered the fuel-switching campaign. Energy conservation schemes required production stops for retrofitting as well as work-routine changes, which were difficult to implement in the stagnated Soviet economy.

63. Castro Ruz, "Discurso pronunciado . . . en la clausura del II Período Ordinario de Sesiones de la Asamblea Nacional del Poder Popular."

64. Naphtha is a fraction of petroleum predominantly used as a feedstock in refineries and to produce high-octane gasoline.

65. Comité Estatal de Estadísticas, *Anuario estadístico de Cuba de 1989*, table XI.9.

66. Pérez-López, "Cuban Oil Reexports," 1.

67. Castro Ruz, "Discurso pronunciado . . . en la clausura del V Congreso de La Federación de Mujeres Cubanas"; Gleijeses, *Visions of Freedom*, 320–21.

68. Coto Acosta, "Primer Fórum Nacional de Energía," 5.

69. *Energía*, "Declaración final del Primer Fórum Nacional de Energía," 17.

70. Castro Ruz, "Discurso pronunciado . . . en el acto de Clausura del Primer Fórum Nacional de Energía."

71. *Revolución*, "Rebajan tarifas eléctricas."

72. Oramas, "La nueva tarifa eléctrica busca el ahorro de petróleo y eliminar el despilfarro de electricidad."

73. Oramas, 5; Oramas, "Entrará en vigor el primero de enero la nueva tarifa eléctrica para centros industriales, de servicios, administrativos y privados."

74. See Wilson, *Everyday Moral Economies*, 6 for a discussion of money as a "token" rather than a measure of exchange value in the Cuban economy.

75. Oramas, "La nueva tarifa eléctrica entra en vigor en octubre," 4.

76. Oramas, "Podrá significar inicialmente el ahorro de 700 millones de KWh—más de 200 mil toneladas de petróleo la aplicación de la nueva tarifa eléctrica."

77. Castro Ruz, "Discurso pronunciado . . . en el Acto Central con Motivo del XXVIII Aniversario del Asalto al Cuartel Moncada."

78. *Energía*, "Guía elaborada por la Comisión Nacional de Energía," 43–48.

79. González Jordan, *Ahorro de energía en Cuba*.

80. Oramas, "Disminuyeron los consumos energéticos en La Habana en el primer trimestre"; Oramas, "Disminuyó en enero el consumo de electricidad en Ciudad de La Habana en 20 542,1 megawatts-hora."

81. Agencia de Información Nacional, "Se registró un sobreconsumo de electricidad en los primeros veinte días de junio."

82. ONE, *Estadísticas energéticas en la Revolución*, tables 14, 67.

83. Castro Ruz, "Discurso pronunciado . . . en el acto de Clausura del Primer Fórum Nacional de Energía."

84. Castro Ruz, "Discurso pronunciado . . . en el Acto Central por el 29 Aniversario del Ataque al Cuartel Moncada."

85. Castro Ruz, "Discurso pronunciado . . . en el acto de Clausura del Primer Fórum Nacional de Energía."

86. See Bérriz Pérez, "Consideraciones sobre el desarrollo histórico del uso de las fuentes renovables de energía, a partir del triunfo de la Revolución Cubana" for a detailed insider history of these efforts, as well as Pons Duarte, *Política energética, política económica y desarrollo*, 41.

87. Castro Ruz, "Discurso pronunciado . . . en el acto de Clausura del Primer Fórum Nacional de Energía."

88. FANJ, *Energía Nuclear*, 1966 (Gaceta Oficial, "30 de Septiembre de 1966, Academia de Ciencias, Resolución No. 15 creando el Grupo de Trabajo de Energía Nuclear").

89. FANJ, *Energía Nuclear*, 1966 (Pentz to Núñez Jiménez, "Report on the Establishment of a Research Institute of Radiation and Nuclear Physics and on a Programme for the Development of Pure and Applied Research in Radiation and Nuclear Physics, for the Training of Scientific and Technical Personnel, and for Laying the Foundations for the Peaceful Application of Atomic Energy to the Industrial and Agricultural Problems of Cuba"; Pentz to Núñez Jiménez, "Propuesta preliminar al Presidente de la Academia de Ciencias de Cuba Capitán Antonio Núñez Jiménez, de un laboratorio de radiación").

90. FANJ, *Energía Nuclear*, 1967 ("Memorándum: Conversación de la Comisión soviética con el Presidente de la República").

91. Castro Ruz, "Discurso pronunciado . . . en el resumen de los actos conmemorativos por el VII Aniversario de la derrota del imperialismo yanki en Playa Girón."

92. Núñez Jiménez, "El acto de inauguración del Instituto de Física Nuclear."

93. Hecht, *Being Nuclear*, 24.

94. Schmid, "Nuclear Colonization?"

95. González Alonso and March Alvarez-Muñoz, "Visión perspectiva del suministro energético nacional," 4.

96. Castro Ruz, "Discurso pronunciado . . . en el acto de Clausura del Primer Fórum Nacional de Energía"; Castro Ruz, "Discurso pronunciado . . . en el Parque 'Céspedes,' de Bayamo."

97. Castro Ruz, "Discurso pronunciado . . . en el acto de Clausura del Primer Fórum Nacional de Energía."

98. Castro Ruz, "Discurso pronunciado . . . en la Inauguración del Combinado Textil de Santiago de Cuba 'Celia Sánchez Manduley'"; Castro Ruz, "Discurso pronunciado . . . en el acto central por el XXXI Aniversario del Asalto al Cuartel Moncada"; see also Castro Díaz-Balart, "La energía nuclear en Cuba."

99. Castro Ruz, "Discurso pronunciado . . . en el acto de Clausura del Primer Fórum Nacional de Energía."

100. Castro Ruz.

101. Schmid, "Nuclear Colonization?"

102. Benjamin-Alvarado, *Power to the People*.

103. Castro Díaz-Balart, *Nuclear Energy*, 291–92.

104. Cederlöf, "The Revolutionary City."

105. Benjamin-Alvarado, *Power to the People*, 76–77.

106. Benjamin-Alvarado, 73.

107. Castro Díaz-Balart, "La energía nuclear en Cuba," 52.

108. Castro Ruz, "Discurso pronunciado . . . en el acto celebrado con motivo de la terminación del montaje de una Unidad en Tallapiedra de la Empresa Eléctrica."

109. Castro Ruz, "Discurso pronunciado . . . en el Parque 'Céspedes,' de Bayamo."

3. BLACKOUT

1. ONE, *Estadísticas energéticas en la Revolución*, tables 65, 67.

2. In another interview, Yuniel, who like Manolito was a transport worker, told me the same story almost exactly, the reference to Nicaragua aside.

3. See Yoss, "What the Russians Left Behind" on Soviet-era nostalgia in contemporary Cuba.

4. Kingsbury, "Combined and Uneven Energy Transitions," 561.

5. *Guerrillero*, "Programación de apagones próxima semana," 7.

6. Rosendahl, *Inside the Revolution*, 169.

7. Presa, *La Revolución Energética*.

8. Monduy Cinteo, "Analizan situación de los combustibles y Programa Energético," 8.

9. Ramos Guadalupe, *Fidel Castro ante los desastres naturales*, 156.

10. Castro Ruz, "Discurso pronunciado . . . en la clausura de la Sesión de Constitución de la Asamblea Nacional, en su cuarta legislatura, y del Consejo de Estado."

11. Deere, "Cuba's National Food Program and Its Prospects for Food Security," 40.

12. Pauslon, "Degrowth," 426.

13. Borowy, "Degrowth and Public Health in Cuba," 19.

14. Boillat, "What Economic Democracy for Degrowth?," 603.

15. Hornborg, *Nature, Society, and Justice in the Anthropocene*, 82–84.

16. de la Fuente, "Recreating Racism"; see also Blue, "The Erosion of Racial Equality in the Context of Cuba's Dual Economy."

17. Shove and Walker, "What Is Energy For?," 46, 55.

18. See, e.g., Lundgren, "Heterosexual Havana"; Rosendahl, *Inside the Revolution*.

19. Forrest, "*Bichos, Maricones* and *Pingueros*," 49–54. As Forrest notably points out, the logo of the FMC portrays "a woman soldier in uniform, with a rifle slung over one shoulder, and holding a baby. Here she becomes also the reproducer, the mother, the nurturer (and vulnerable with it), as well as a fighter and a heroine" (50).

20. Josée Davidson and Krull, "Adapting to Cuba's Shifting Food Landscapes," 62.

21. See the foundational work by Vayda, "Progressive Contextualization" and Blaikie and Brookfield, *Land Degradation and Society* on explanation and contextualization in political ecology.

22. Castro Ruz, "Discurso pronunciado . . . en el Acto Central por el XXX Aniversario de los Comités de Defensa de la Revolución."

23. Jiménez García, "Energía," 23.

24. *Guerrillero*, "Programación de apagones próxima semana," 7; *Guerrillero*, "Informan sobre reajustes en la programación de apagones," 7.

25. Castro Ruz, "Discurso pronunciado . . . en el Acto Central por el XXX Aniversario de los Comités de Defensa de la Revolución"; Castro Ruz, "Discurso pronunciado . . . en la Inauguración del IV Congreso del Partido Comunista de Cuba"; ONE, *Estadísticas energéticas en la Revolución*, tables 14, 25.

26. Castro Ruz, "Discurso pronunciado . . . en la Inauguración del IV Congreso del Partido Comunista de Cuba"; Castro Ruz, "Discurso pronunciado . . . en el acto por el XXXIX Aniversario del Asalto al Cuartel Moncada y el XXXV del Levantamiento de Cienfuegos."

27. Castro Ruz, "Discurso pronunciado . . . en la Inauguración del IV Congreso del Partido Comunista de Cuba."

28. See, e.g., Alonso-Pippo et al., "Sugarcane Energy Use"; Mesa-Lago, "Assessing Economic and Social Performance in the Cuban Transition of the 1990s"; Pérez-López, *The Economics of Cuban Sugar*.

29. It is of course still questionable to what extent trade between Cuba and the Soviet Union was conducted on equal terms, the trade agreements resulting from negotiations between an agrarian island-nation and an industrialized superpower.

30. Castro Ruz, "Discurso pronunciado . . . en la Inauguración del IV Congreso del Partido Comunista de Cuba."

31. Emmanuel, *Unequal Exchange*.

32. Castro Ruz, "Discurso pronunciado . . . en el Acto Central por el XXX Aniversario de los Comités de Defensa de la Revolución."

33. Castro Ruz, "Discurso pronunciado . . . en la Inauguración Oficial de la Fábrica de Cemento 'Carlos Marx.'"

34. Castro Ruz.

35. Comité Estatal de Estadísticas, *Anuario estadístico de Cuba de 1989*, table XI.9. Fidel reported on the level of oil re-exports to the Third Congress of the PCC, Castro Ruz, "Discurso pronunciado . . . en la clausura de la Sesión diferida del Tercer Congreso del Partido Comunista de Cuba."

36. Pérez-Stable, *The Cuban Revolution*, 125; Eckstein, "The Rectification of Errors or the Errors of the Rectification Process in Cuba?"

37. Bain, "Soviet/Cuba Relations 1985–1991."

38. Bain; Shearman, "Gorbachev and the Restructuring of Soviet-Cuban Relations."

39. Mesa-Lago, "Assessing Economic and Social Performance in the Cuban Transition of the 1990s," 868; Castro Ruz, "Discurso pronunciado . . . en el acto por el XXXIX Aniversario del Asalto al Cuartel Moncada y el XXXV del Levantamiento de Cienfuegos"; ONE, *Estadísticas energéticas en la Revolución*, tables 14, 25.

40. Castro Ruz, "Discurso pronunciado . . . en el acto por el XXXIX Aniversario del Asalto al Cuartel Moncada y el XXXV del Levantamiento de Cienfuegos." See also Castro Ruz, "Discurso pronunciado . . . en el Acto Central por el XXX Aniversario de los Comités de Defensa de la Revolución"; Castro Ruz, "Discurso

pronunciado . . . en la Inauguración del IV Congreso del Partido Comunista de Cuba."

41. Hornborg, "Zero-Sum World," 242.

42. Högselius, *Red Gas*, 204–9.

43. Mehrotra, *India and the Soviet Union*, 173.

44. Castro Ruz, "Discurso pronunciado . . . en la clausura del acto central por el XL Aniversario del Asalto a los Cuarteles Moncada y 'Carlos Manuel de Céspedes.'"

45. *Granma*, "Reseña de los debates," 3.

46. Antúnez, "Los eléctricos no sólo dan apagones," 1; Antúnez, "Hacer en casa lo que antes costaba divisas."

47. Walker, *Energy and Rhythm*, 96.

48. Rodríguez Castellón, "La industria eléctrica cubana," 158; Presa, *La Revolución Energética*, slide 25; ONEI, *Anuario estadístico de Cuba 2013*, table 10.19.

49. MINBAS, *Ahorro de energía*, 46. These figures are my approximations from a printed graph.

50. MINBAS.

51. This assessment is based on a review of approximately fifteen timetables in *Guerrillero* from 1993.

52. Rosendahl, *Inside the Revolution*, 37–38; Cabrera Arús, "The Material Promise of Socialist Modernity," 203–5.

53. Castro Ruz, "Discurso pronunciado . . . en la Inauguración del IV Congreso del Partido Comunista de Cuba."

54. Castro Ruz, "Discurso pronunciado . . . en la Clausura del V Congreso de la Federación de Mujeres Cubanas," emphasis added.

55. Castro Ruz, "Discurso pronunciado . . . en el Acto Central por el XXX Aniversario de los Comités de Defensa de la Revolución."

56. Powell, "Neoliberalism, the Special Period and Solidarity in Cuba," 180; Hernández-Reguant, "Writing the Special Period," 2.

57. Castro Ruz, "Discurso pronunciado . . . en el Acto Central por el XXX Aniversario de los Comités de Defensa de la Revolución."

58. Castro Ruz, "Discurso pronunciado . . . en la Clausura del V Congreso de la Federación de Mujeres Cubanas."

59. Koselleck, *Futures Past*.

60. Castro Ruz, "Discurso pronunciado . . . en el acto por el XXXIX Aniversario del Asalto al Cuartel Moncada y el XXXV del Levantamiento de Cienfuegos"; see also Castro Díaz-Balart, *Nuclear Energy*, 314.

61. Cuban-Soviet trade had previously taken place under CIF conditions ("cost, insurance, freight") whereby the Soviet Union assumed responsibility for all cargo until it was delivered. Russia instead demanded FOB conditions ("free on board"), which would relieve them of responsibility as soon as cargo was loaded on to a ship.

62. Castro Ruz, "Discurso pronunciado . . . en el acto por el XXXIX Aniversario del Asalto al Cuartel Moncada y el XXXV del Levantamiento de Cienfuegos." The Cuban government decided to fully cancel the project in 2000 during a Russian state visit to Havana.

63. Castro Ruz, "Discurso pronunciado . . . en la Clausura del V Congreso de la Federación de Mujeres Cubanas," emphases added.

64. See, e.g., Gordy, "'Sales + Economy + Efficiency = Revolution'?"; Padrón Hernández, "Beans and Roses," 139–43; Wilson, *Everyday Moral Economies.*

65. Powell, "Neoliberalism, the Special Period and Solidarity in Cuba"; Forrest, "*Bichos, Maricones* and *Pingueros*"; see also de la Fuente, "Recreating Racism."

66. Ávila, "Regalao murió en el 80." For copyright reasons, I am unable to reproduce the exact lyrics here.

67. Wilson, *Everyday Moral Economies*, 86.

68. See also Kapcia, *Cuba in Revolution*, 85 on this point.

4. SOCIALIST REDISTRIBUTION AND AUTONOMOUS INFRASTRUCTURE

1. Calder, *The New Continentalism*, 53; see also Aklin and Urpelainen, *Renewables.*

2. Villaescusa, "Asambleas sindicales para ahorrar energía"; *Guerrillero*, "El Programa Energético Integral"; Lee, "El vital tema de la energía, desde la base."

3. López, "Convocan a concurso sobre ahorro de energía," 7.

4. Blanco, "Primer Taller Científico de Energía."

5. Villaescusa, "Asambleas sindicales para ahorrar energía," 1; *Granma*, "Reseña de los debates."

6. *Granma*, "Reseña de los debates," 2.

7. Lee, "El vital tema de la energía, desde la base," 5.

8. Rodríguez Martínez, "La energía en Cuba"; see also *Granma*, "Reseña de los debates"; Altshuler et al., *Suplemento especial.*

9. Huber and McCarthy, "Beyond the Subterranean Energy Regime?"; Sieferle, *The Subterranean Forest*; Wrigley, *The Path to Sustained Growth.*

10. See, e.g., *Cuba Foreign Trade*, "The Cuban Basic Industry."

11. ONEI, *Anuario estadístico de Cuba 2013*, table 10.04. In January 2015, Cuba had 124 million barrels of proven crude oil reserves. Exploration attempts by major companies including CNPC, Petrobras, Petronas, PDVSA, Repsol, Rosneft, and Zarubezhneft had not yielded any significant results in the deepwater offshore, and exploration had shifted onshore, primarily to the provinces of Mayabeque and Matanzas, Sancti Spíritus and Ciego de Ávila; see EIA, "Cuba."

12. Antúnez, "Hacer en casa lo que antes costaba divisas," 1.

13. Benjamin-Alvarado, *Power to the People*, 96; Camacho, "Interviene Francia en modernización de centrales eléctricas"; see also Feinberg, "Foreign Investment in the New Cuban Economy" on the difficulty of assessing foreign investment data in Cuba.

14. Jiménez García, "Energía," 25; Torres-Martínez, "Nuevo Sistema Electroenergético Nacional en Cuba, basado fundamentalmente en biomasa cañera," 324.

15. Cocq and McDonald, "Minding the Undertow"; Pérez-López, "Cuba's Thrust to Attract Foreign Investment."

16. Werniuk, "Marching to a Different Drum."

17. In combined cycle gas turbine (CCGT) plants such as these, natural gas is first combusted in a turbine whose mechanical energy is used to generate electricity. The hot exhaust gases are then captured and used to produce steam. The steam is led into a second turbine to generate even more electricity. Thus, two thermodynamic cycles are combined, utilizing only one heat source to achieve a significantly higher conversion efficiency than in a single cycle.

18. Hirsh, *Power Loss*.

19. *Granma*, "Reseña de los debates," 2.

20. *Granma*; Rodríguez Martínez, "La energía en Cuba," 7.

21. Alonso-Pippo et al., "Sugarcane Energy Use."

22. Gómez, "Resolver los problemas de la población con el concurso de todos y concretar en cada sitio el Programa Energético," 8.

23. *Granma*, "Se está discutiendo el tema de la energía en medio de las condiciones más complejas."

24. See further Torres-Martínez, "Nuevo Sistema Electroenergético Nacional en Cuba, basado fundamentalmente en biomasa cañera," 324.

25. Alonso-Pippo et al., "Sugarcane Energy Use," 2165.

26. Hornborg et al., "Has Cuba Exposed the Myth of 'Free' Solar Power?," 990. On investments in renewables, see Bérriz and Madruga, *Cuba y las fuentes renovables de energía*; Suárez Rodríguez and Beatón Soler, "Estado y perspectivas de las energías renovables en Cuba"; and Cherni and Hill, "Energy and Policy Providing for Sustainable Livelihoods in Remote Locations."

27. *Guerrillero*, "El Programa Energético Integral," 8. The media statement gives no indication of the ecological composition of these energy forests.

28. López and Avila, "Acuerdan asambleas municipales del Poder Popular medidas para concretar ahorro de energético y perfeccionar los consejos."

29. *Granma*, "Reseña de los debates," 3.

30. Crespo Porbén, "Valoran principales cuadros comportamiento del Programa Energético."

31. Gómez, "Resolver los problemas de la población con el concurso de todos y concretar en cada sitio el Programa Energético," 8.

32. Borrego, "Que la paja de arroz no vaya a los hornos de desperdicio."

33. Crespo Porbén, "Valoran principales cuadros comportamiento del Programa Energético."

34. Rodríguez Martínez, "La energía en Cuba," 7; Terrero, "¿Cuántas patas tiene el gato?," 31.

35. Smil, *Power Density*, 237.

36. Huber and McCarthy, "Beyond the Subterranean Energy Regime?"; "Hornborg et al., "Has Cuba Exposed the Myth of 'Free' Solar Power?"; Scheidel and Sorman, "Energy Transitions and the Global Land Rush."

37. Crespo Porbén, "Vivir de la candela," 4.

38. Guevara, "Cuba," 133.

39. ONEI, *Anuario estadístico de Cuba 2015*, table 10.14.

40. López and Avila, "Acuerdan asambleas municipales del Poder Popular medidas para concretar ahorro de energético y perfeccionar los consejos," 7.

41. Gómez, "Resolver los problemas de la población con el concurso de todos y concretar en cada sitio el Programa Energético," 8.

42. Deere, "Cuba's National Food Program and Its Prospects for Food Security"; Enríquez, "Economic Reform and Repeasantization in Post-1990 Cuba"; Powell, "Neoliberalism, the Special Period and Solidarity in Cuba."

43. *Guerrillero*, "El Programa Energético Integral," 8; see also, e.g., López and Avila, "Acuerdan asambleas municipales del Poder Popular medidas para concretar ahorro de energético y perfeccionar los consejos," 7.

44. Altshuler et al., *Suplemento especial*, 30.

45. Altshuler et al., 30–31; Arrastía Ávila, "Preguntas y respuestas sobre el ahorro de energía eléctrica."

46. MINBAS, *Ahorro de energía y respeto ambiental*, xiii; MINBAS, *Ahorro de energía*; see further Stricker, *Toward a Culture of Nature*.

47. Nicado, "¿Cómo 'estirar' el keroseno?," 3.

48. de la Cruz, "'Moñito' ahorrativo," 3.

49. Crespo Porbén, "Valoran principales cuadros comportamiento del Programa Energético."

50. Gómez, "Resolver los problemas de la población con el concurso de todos y concretar en cada sitio el Programa Energético"; Lee, "El vital tema de la energía, desde la base." See Faris et al., "Effects of Magnetic Field on Fuel Consumption and Exhaust Emissions in Two-Stroke Engines" for a further, more technical explanation of fuel magnetization.

51. Gómez, "Resolver los problemas de la población con el concurso de todos y concretar en cada sitio el Programa Energético," 8. MINAG in Pinar del Río reported that, by the end of March 1993, there were 240 windmills in use in the province and another 50 under construction, see Crespo Porbén, "Valoran principales cuadros comportamiento del Programa Energético" and Rodríguez Martínez, "La energía en Cuba," 7.

52. Crespo Porbén, "Valoran principales cuadros comportamiento del Programa Energético," 8; *Guerrillero*, "El Programa Energético Integral."

53. *Granma*, "Reseña de los debates," 2.

54. See Bérriz Pérez, "Calentador solar de tubos al vacío," for a review of Cuban solar heater technologies.

55. Bérriz and Madruga, *Cuba y las fuentes renovables de energía*, 14; Turrini, *El camino del sol*, 148–50.

56. Suárez Rodríguez and Beatón Soler, "Estado y perspectivas de las energías renovables en Cuba."

57. *Guerrillero*, "El Programa Energético Integral," 8; López and Avila, "Acuerdan asambleas municipales del Poder Popular medidas para concretar ahorro de energético y perfeccionar los consejos."

58. Gómez, "Resolver los problemas de la población con el concurso de todos y concretar en cada sitio el Programa Energético," 8; see also López and Avila, "Acuerdan asambleas municipales del Poder Popular medidas para concretar ahorro de energético y perfeccionar los consejos."

59. Crespo Porbén, "Valoran principales cuadros comportamiento del Programa Energético," 8.

60. *Granma*, "Reseña de los debates," 3.

61. Ríos and Ponce, "Mechanization, Animal Traction, and Sustainable Agriculture," 155. These figures differ slightly from those reported in CMEA Secretariat, *Statistical Yearbook of Member States of the Council for Mutual Economic Assistance 1979*, 274.

62. Enoch et al., "The Effect of Economic Restrictions on Transport Practices in Cuba."

63. Hernández-Reguant, "Writing the Special Period," 3; see also Borowy, "Degrowth and Public Health in Cuba."

64. Bérriz Pérez, "Consideraciones sobre el desarrollo histórico del uso de las fuentes renovables de energía, a partir del triunfo de la Revolución Cubana," 50.

65. Pagés Díaz, "Potencialidad de residuos agroindustriales para producir biogás."

66. Anusas and Ingold, "The Charge against Electricity," 542.

67. Meehan, "Tool-Power," 217.

68. Dove and Kammen, *Science, Society and the Environment*.

69. Dove and Kammen, 123.

70. Altieri, *Agroecology*; Cederlöf, "Low-Carbon Food Supply."

71. Huber, "Hidden Abodes."

72. Dove and Kammen, *Science, Society and the Environment*, 81.

73. Dove and Kammen, 86–87.

74. Dove and Kammen, 78; see also Maring, "The Strategy of Shifting Cultivators in West Kalimantan in Adapting to the Market Economy."

75. Engels, "On Authority."

76. Scheer, *Energy Autonomy*, 239–40.

77. Scheer, *The Solar Economy*, 89; see also, e.g., Winner, *The Whale and the Reactor* and in particular the chapter "Do Artifacts Have Politics?" for a fuller discussion.

78. Lapaz and Bérriz, "Energía solar." In the context of Puerto Rico, Smith-Nonini, "The Debt/Energy Nexus behind Puerto Rico's Long Blackout," 79, documents a similar argument being made about solar energy and popular control in the aftermath of the cataclysmic hurricane María (2018).

79. Mitchell, *Carbon Democracy*, 266–67.

80. Massey, *World City*, 167.

81. CMEA Secretariat, *Statistical Yearbook of the Member States of the Council for Mutual Economic Assistance 1979*, 274.

5. THE ENERGY REVOLUTION

1. Mayoral, "Detallada información al pueblo sobre los problemas en el servicio eléctrico"; Mayoral and de la Osa, "La población ha sido, es y será lo más sagrado para la Revolución."

2. León Moya, "Continúa el trabajo para restablecer la estabilidad"; Mayoral, "Detallada información al pueblo sobre los problemas en el servicio eléctrico."

3. Tamayo León and Alemany Gutiérrez, "Tenemos los recursos para recuperarnos"; Suárez, "Batalla de alta tensión"; Regalado and Alemany, "La electricidad se restablece."

4. Josée Davidson and Krull, "Adapting to Cuba's Shifting Food Landscapes."

5. Cardoso and Brizuela, "Continúan las medidas alternativas," 1; see also Molina Pérez, "Reconoce Lage esfuerzo de los pinareños."

6. Regalado, "Se activan pozos con bombas de mano"; Regalado, "Permanecerán activas plantas eléctricas."

7. Suárez, "Agua desde la refinería del 'Martí.'"

8. Mayoral and de la Osa, "La población ha sido, es y será lo más sagrado para la Revolución," 1.

9. From 0.19 to 0.13 kg CO_2 per USD_{2010} and from 0.09 to 0.05 toe per thousand USD_{2010} between 2003 and 2009; IEA, "Country Data."

10. Castro Ruz, "Discurso pronunciado . . . en el acto por el aniversario 60 de su ingreso a la universidad."

11. Viego Felipe et al., *Temas especiales de sistemas eléctricos industriales*, 10.

12. Mayoral, "Revolución Energética con resultados loables"; Monteagudo and Terrero, "En Cuba Revolución Energética," 31; ONEI, *Anuario estadístico de Cuba 2014*, table 10.18. In 2006, UNE also installed 120 MW in isolated systems, on keys and other geographically remote areas, as a way to further electrify the country.

13. Monteagudo and Terrero, "En Cuba Revolución Energética," 30.

14. Monteagudo, "Paso Real no cree en apagones," 21.

15. Castro Ruz, "Discurso pronunciado . . . en el acto por el Día Internacional de los Trabajadores." In a speech earlier in 2006, Fidel Castro stated that the plan was to install 4,158 emergency generators, Castro Ruz, "Discurso pronunciado

. . . en ocasión del aniversario 47 de su entrada en Pinar del Río, el acto por la culminación del montaje de los grupos electrógenos en esa provincia." In 2007, *Granma* reported that 6,841 emergency generators had been installed. Mayoral, "Revolución Energética con resultados loables."

16. 4.40 of 17.73 TWh; see ONEI, *Anuario estadístico de Cuba 2016*, table 10.16.

17. Viego Felipe et al., *Temas especiales de sistemas eléctricos industriales*, 12.

18. Moreno Figueredo, *Energía eólica*, 31.

19. Harrison and Popke, "Geographies of Energy Transition in the Caribbean"; Shirley and Kammen, "Renewable Energy Sector Development in the Caribbean"; Smith-Nonini, "The Debt/Energy Nexus behind Puerto Rico's Long Blackout."

20. ONEI, *Anuario estadístico de Cuba 2016*, table 10.16.

21. Castro Ruz, "Discurso pronunciado . . . en el acto por el Día Internacional de los Trabajadores."

22. Monteagudo and Terrero, "En Cuba Revolución Energética," 32.

23. Lage Dávila, "Puede decirse que en menos de tres años el país alcanzó una capacidad de generación eléctrica suficientemente por encima de su demanda máxima"; see also Monteagudo and Terrero, 33.

24. Castro Ruz, "Discurso pronunciado . . . en el acto por el Día Internacional de los Trabajadores."

25. Castro Ruz, "Discurso pronunciado . . . en ocasión del aniversario 47 de su entrada en Pinar del Río, el acto por la culminación del montaje de los grupos electrógenos en esa provincia."

26. Monteagudo, "Paso Real no cree en apagones," 21.

27. First figure from Monteagudo; quote from Castro Ruz, "Discurso pronunciado . . . en el acto por el aniversario 60 de su ingreso a la universidad."

28. Mayoral, "Revolución Energética con resultados loables."

29. Castro Ruz, "Discurso pronunciado . . . en el acto por el Día Internacional de los Trabajadores."

30. Lage Dávila, "Puede decirse que en menos de tres años el país alcanzó una capacidad de generación eléctrica suficientemente por encima de su demanda máxima."

31. *Juventud Rebelde*, "El día después del paso de Gustav"; *Juventud Rebelde*, "Restablecen paulatinamente servicio eléctrico en la Isla de la Juventud"; personal communications.

32. Monteagudo, "Paso Real no cree en apagones," 21.

33. Barnes, "States of Maintenance," 7. In another notable study, Zimmerer, "Rescaling Irrigation in Latin America" showcases a similar dynamic in the case of the Inca state's canal-based irrigation systems at the time of colonial disruption.

34. Coronil, *The Magical State*, 89.

35. Kingsbury, "Oil's Colonial Residues"; Mommer, "Subversive Oil"; Wilpert, *Changing Venezuela by Taking Power*.

36. Cederlöf and Kingsbury, "On PetroCaribe," 127.

37. Gleijeses, *Visions of Freedom*.

38. Azicri, "The Castro-Chávez Alliance"; Yaffe, "The Bolivarian Alliance for the Americas."

39. PetroCaribe, "Acuerdo de Cooperación Energética Petrocaribe." See Cederlöf and Kingsbury, "On PetroCaribe" for an extensive discussion of the political rationale behind PetroCaribe and its production as a territorial-infrastructural space.

40. Harrison and Popke, "Reassembling Caribbean Energy?," 221–22.

41. PetroCaribe, *PetroCaribe Management Report Quarter I*, 21.

42. Cederlöf and Kingsbury, "On PetroCaribe," 131; Kirk, "Cuban Medical Cooperation within ALBA."

43. *Agencia Bolivariana de Noticias*, "Chávez"; also quoted in Kirk, "Cuban Medical Cooperation within ALBA," 232.

44. Cederlöf and Kingsbury, "On PetroCaribe," 129.

45. Chávez Frías, "Petrocaribe es parte del cambio de época que estamos viviendo"; also quoted in Cederlöf and Kingsbury, 130.

46. MPetroMin, *Petróleo y otros datos estadísticos* (54 ed.); table 67; MPetroMin, *Petróleo y otros datos estadísticos* (55 ed.), table 67.

47. The conversion from barrels of oil (volume) to metric tons (weight) is not straightforward, as it depends on the oil's density. Here I have used OPEC's conversion factor for Venezuelan crude of 6.68816 barrels per ton, see OPEC, *Annual Statistical Bulletin 2016*, 124. However, the dataset from MPetroMin also includes oil derivatives, which have substantially lower densities than Venezuela's generally very heavy crude. The conversion therefore indicates the highest possible weight of trade, granted that crude is the heaviest form of oil per unit of volume. In reality, the figure was lower, as Cuba also imported various distillates from Venezuela.

48. PetroCaribe, *PetroCaribe Management Report Quarter I*, table 1.

49. Mayoral, "Revolución Energética con resultados loables," 3; Castro Ruz, "Medidas sobre las tarifas eléctricas, los incrementos salariales y los de la Seguridad y la Asistencia Social"; Monteagudo and Terrero, "En Cuba Revolución Energética."

50. Castro Ruz, "Discurso pronunciado . . . en el acto por el aniversario 60 de su ingreso a la universidad"; Arrastía Ávila, "Preguntas y respuestas sobre el ahorro de energía eléctrica," 82.

51. Castro Ruz, "Discurso pronunciado . . . en la entrega de 101 vehículos a la Unión Eléctrica."

52. Knuth, "Cities and Planetary Repair," 493.

53. Knuth, 493; see also Thoyre, "Negawatt Resource Frontiers."

54. Cheng, "Fidel Castro and 'China's Lesson for Cuba,'" 26.

55. Cheng, 26; Hearn, "China, Global Governance and the Future of Cuba"; Ratliff, "Cuban Foreign Policy toward Far East and Southeast Asia."

56. León Moya, "Continúa el trabajo para restablecer la estabilidad," 3.

57. DURE, *La energía y mi país.*

58. Pérez et al., "La Revolución da un paso hacia su invulnerabilidad política," 4.

59. *Guerrillero*, "Sobre la asignación del combustible doméstico para reserva de cada familia pinareña," 2.

60. Font, "Cuba and Castro"; Kapcia, "Educational Revolution and Revolutionary Morality in Cuba"; Yaffe, *We Are Cuba!*

61. Kapcia, "Educational Revolution and Revolutionary Morality in Cuba."

62. Hansing, "Changes from Below," 18–19.

63. Kapcia, "Educational Revolution and Revolutionary Morality in Cuba," 404.

64. Guerra, *Visions of Power in Cuba*, 5. See also Rojas, "The New Text of the Revolution" on debates on the historicization of "the Revolution" in Cuba, which showcases a plurality of ideas on this topic.

65. Castro Ruz, "Discurso pronunciado . . . en el acto por el Día Internacional de los Trabajadores."

66. Barry, "Thermodynamics, Matter, Politics," 122.

67. Stengers, *Cosmopolitics I*, 210 quoted in Barry, 117.

68. Cederlöf and Hornborg, "System Boundaries as Epistemological and Ethnographic Problems," 115.

69. Daggett, *The Birth of Energy.*

70. Daggett, 133.

71. This is based on a review of twenty-three MSc dissertations from CEETES at the University of Pinar del Río and two introductory lectures to the dissertation projects that I attended in 2015.

72. Pedrera, "Gestión energética en Empresa Eléctrica de Pinar del Río," 9.

73. Sánchez Almeira, "Gestión energética empresarial," 13.

74. Suárez Rivas, "Kilowatts en saco roto."

75. Alemany Gutiérrez, "Incumplidores y excedidos en planes de electricidad."

76. Suárez Rivas, "Kilowatts en saco roto."

77. Suárez Rivas, "Desmesurado gasto eléctrico por los organismos en Pinar del Río," 3.

78. Monteagudo and Terrero, "En Cuba Revolución Energética," 34.

79. Moreno Figueredo, *Energía eólica*, 30–32.

80. Castro Ruz, "Discurso pronunciado . . . en el acto por el Día Internacional de los Trabajadores."

81. Castro Ruz, "Discurso pronunciado . . . en la entrega de 101 vehículos a la Unión Eléctrica"; see also, e.g., *Bohemia*, "Revolución energética en línea."

82. Castro Ruz, "Discurso pronunciado . . . en ocasión del aniversario 47 de su entrada en Pinar del Río, el acto por la culminación del montaje de los grupos electrógenos en esa provincia."

83. Alemany Gutiérrez, "La energía de una Revolución," 2.

84. Alemany Gutiérrez; Castro Ruz, "Discurso pronunciado . . . en el acto por el aniversario 60 de su ingreso a la universidad"; Arrastía-Ávila and Glidden, "Cuba's Energy Revolution and 2030 Policy Goals."

85. Castro Ruz, "Discurso del Comandante en Jefe . . . durante su visita al Grupo Empresarial Panda, Nanjing"; EIE, "ATEC Systems." The Cuban government later also formed a joint venture with the Chinese electronics manufacturer Haier to assemble an ATEC-Haier television set in Cuba. Haier notably also delivered refrigerators to Cuba.

86. Castro Ruz, "Discurso pronunciado . . . en el acto por el aniversario 60 de su ingreso a la universidad."

87. Castro Ruz, "Discurso pronunciado . . . en ocasión del aniversario 47 de su entrada en Pinar del Río, el acto por la culminación del montaje de los grupos electrógenos en esa provincia."

88. E.g., Castro Ruz, "Discurso pronunciado . . . en la entrega de 101 vehículos a la Unión Eléctrica"; Castro Ruz, "Discurso pronunciado . . . en el acto por el Día Internacional de los Trabajadores."

89. Monteagudo and Terrero, "En Cuba Revolución Energética," 34.

90. Josée Davidson and Krull, "Adapting to Cuba's Shifting Food Landscapes."

91. Sartorio, "El amor entró en la cocina," 8.

92. Castro Ruz, "Discurso pronunciado . . . en el acto por el Día Internacional de los Trabajadores."

93. Wilson, *Everyday Moral Economies*, 106–108.

94. Thoyre, "Negawatt Resource Frontiers."

95. E.g., Pérez et al., "La Revolución da un paso hacia su invulnerabilidad política."

96. Castro Ruz, "Discurso pronunciado . . . en ocasión del aniversario 47 de su entrada en Pinar del Río, el acto por la culminación del montaje de los grupos electrógenos en esa provincia."

97. *Guerrillero*, "Sobre la asignación del combustible doméstico para reserva de cada familia pinareña"; personal communications.

98. Households in the older parts of Havana were an exception, as they had access to gas for a fixed, heavily subsidized monthly fee through a prerevolutionary gas infrastructure. These households were nonetheless also offered the electrical appliances.

99. ONEI, *Anuario estadístico de Cuba 2016*, table 10.13. The reliability of these statistics is questionable as they likely fail to account for informally traded liquid fuels.

100. Mitchell, *Carbon Democracy*.

101. Castro Ruz, "Discurso pronunciado . . . en el acto por el aniversario 60 de su ingreso a la universidad."

102. Castro Ruz.

103. Castro Ruz, "Medidas sobre las tarifas eléctricas, los incrementos salariales y los de la Seguridad y la Asistencia Social."

104. Bouzarovski et al., "Multiple Transformations," 25.

105. Raúl Castro quoted in Bobes, "Reformas en Cuba," 166; see further Torres Pérez, "Updating the Cuban Economy," 262.

CONCLUSION: ENERGY TRANSITIONS AND INFRASTRUCTURAL FORM

1. Sánchez and Sánchez Serra, "Fuentes renovables de energía," 4.

2. The much-publicized rapprochement between Cuba and the United States under the Obama administration was quickly reversed by President Trump after which Cuba faced new direct as well as indirect sanctions, the latter via its trade with Venezuela; see LeoGrande, "Reversing the Irreversible."

3. ONEI, *Anuario estadístico de Cuba 2018*, table 10.6; ONEI, *Anuario estadístico de Cuba 2020*, table 10.6. See Cederlöf and Kingsbury, "On Petro-Caribe" and Kingsbury, "Combined and Uneven Energy Transitions" for a fuller discussion.

4. Castro Morales and Suárez Rivas, "Otra mala noticia para el imperio."

5. Martínez García, "Combustible, problema sin resolver."

6. See Fletcher, "La Revolución Energética" for a critical discussion of Cuba's dependence on Venezuelan oil and an analysis on how it may be reduced. See notably Fischhendler et al., "Light at the End of the Panel" and Overland, "The Geopolitics of Renewable Energy" on how renewables currently are reworking geopolitical relations in an international context.

7. *Juventud Rebelde*, "Cuba cuenta con un nuevo parque solar fotovoltaico." On the international political economy of Cuba's solar energy plans, see Hornborg et al., "Has Cuba Exposed the Myth of 'Free' Solar Power?"

8. Funes Monzote, *From Rainforest to Cane Field in Cuba*, 328n14; Travieso Pedroso and Kaltschmitt, "*Dichrostachys cinerea* as Possible Energy Crop."

9. García Elisalde, "La energía que renueva a Cuba." Silver, "Disrupted Infrastructures," similarly finds that solar panel access in Accra, Ghana, mainly was an option for the urban middle class, creating an unequal urban energy landscape.

10. E.g., Kallis et al., "Research on Degrowth"; Urry, *Societies beyond Oil*; WWF, *Living Planet Report*.

11. Heffron and McCauley, "The Concept of Energy Justice across the Disciplines"; Jenkins et al., "Energy Justice"; Sovacool and Dworkin, "Energy Justice."

12. Bridge et al., "Energy Infrastructure and the Fate of the Nation"; Schwenkel, "The Current Never Stops"; see also Swyngedouw, *Liquid Power*.

13. Marx, *Capital*, 133.

14. The same was true for example in the United States, see Harrison, "The Historical-Geographical Construction of Power" and Hirsh, *Power Loss.*

15. Borowy, "Degrowth and Public Health in Cuba," 18.

16. Bustamante and Lambe, "Cuba's Revolution from Within."

17. Barca, "The Labor(s) of Degrowth," 208.

18. Pons Duarte, *Política energética, política económica y desarrollo.*

19. Bridge et al., "Geographies of Energy Transition"; Castán Broto, *Urban Energy Landscapes*; Huber, *Lifeblood*; Malm, *Fossil Capital.*

20. Cederlöf, "Out of Steam"; Shove and Walker, "What Is Energy For?"

21. On a "whole systems" approach, see also, e.g., Blondeel et al., "The Geopolitics of Energy System Transformation" and Jenkins et al., "Energy Justice."

22. Anusas and Ingold, "The Charge against Electricity," 549.

23. Mann, "The Autonomous Power of the State; Harris, "State as Socionatural Effect"; Meehan, "Tool-Power"; Power and Kirshner, "Powering the State."

24. Boyer, "Energopower"; Mitchell, *Carbon Democracy.*

25. Barnes, "States of Maintenance"; Cederlöf, "Maintaining Power."

26. Hornborg et al., "Has Cuba Exposed the Myth of 'Free' Solar Power?," 993.

27. Knuth et al., "Rethinking Climate Futures through Urban Fabrics"; Malm, *Fossil Capital*; Mitchell, *Carbon Democracy.*

28. See Guzón Camporredondo, *Desarrollo local en Cuba* on locally based development planning, which closely ties into the work of organizations such as CubaSolar and Fundación "Antonio Núñez Jiménez" de la Naturaleza y el Hombre.

29. See further Castán Broto et al., "Energy Justice and Sustainability Transitions in Mozambique."

30. See, e.g., Aronoff et al., *A Planet to Win*; Huber, "Fossilized Liberation"; and Robbins, "Is Less More . . . or Is More Less?"

31. Gómez-Baggethun, "More Is More"; Kallis, *Limits*; Paulson, "Degrowth."

32. Boyer, "Infrastructure, Potential Energy, Revolution," 228.

33. Huber, "Fossilized Liberation," 512.

34. Schwartzman, "Solar Communism," 320.

35. Boyer, "Infrastructure, Potential Energy, Revolution," 237.

36. Funes Monzote, *"Geotransformación."*

37. See also Hornborg, *Global Magic.*

38. Hornborg; Hornborg, *Nature, Society, and Justice in the Anthropocene.*

39. Bridge, "The Map Is Not the Territory," 16; Lennon, "Decolonizing Energy"; Newell, "Race and the Politics of Energy Transitions."

40. Núñez Jiménez, *Geografía de Cuba*, 24.

41. Harrison and Popke, "Geographies of Renewable Energy Transition in the Caribbean," 165; see also Clark, "Island Development" and Malm, "No Island Is an 'Island'."

42. Massey, "Power-Geometry and a Progressive Sense of Place."

43. On the politics of connectivity, see Clark and Clark, "Isolating Connections—Connecting Isolations."

44. Wright, *Sustainable Agriculture and Food Security in an Era of Oil Scarcity*; Rosset and Val, "The 'Campesino a Campesino' Agroecology Movement in Cuba."

Bibliography

POLITICAL SPEECHES

Unless otherwise indicated, all speeches by Fidel Castro Ruz are the official transcripts of the Cuban Council of State accessed in the original Spanish at http://www.cuba.cu/gobierno/discursos.

Castro Ruz, Fidel. 1960. "Discurso pronunciado . . . en la Asamblea General de los Trabajadores de Plantas Eléctricas." December 14, Teatro de la CTC, Havana.

———. 1963. "Total la mecanización de nuestra zafra en 2 años." *Revolución*, 2nd ed., June 5: 1–6.

———. 1968. "Discurso pronunciado . . . en el resumen de los actos conmemorativos por el VII Aniversario de la derrota del imperialismo yanki en Playa Girón." April 19, Playa Girón.

———. 1972. "Discurso pronunciado . . . en el acto celebrado con motivo de la terminación del montaje de una Unidad en Tallapiedra de la Empresa Eléctrica." July 23, Tallapiedra, Havana.

———. 1974. "Discurso pronunciado . . . en la concentración popular efectuada en la Plaza de la Revolución 'José Martí', en Honor del compañero Leonid Ilich Brezhnev, Secretario General del Comité Central del Partido Comunista de la Unión Soviética, y la delegación que lo acompaña." January 29, Plaza de la Revolución, Havana.

———. 1978. "Discurso pronunciado . . . de Inauguración del Segundo Bloque de 100 Megawatts de la Termoeléctrica 'Máximo Gómez', de Mariel, La Habana." February 15, Mariel.

———. 1978. "Discurso pronunciado . . . en el acto de Inauguración de la Termoeléctrica 'Carlos Manuel de Céspedes', celebrado en Ocasión de Conmemorarse el Día del Constructor." December 5, Cienfuegos.

———. 1980. "Discurso pronunciado . . . en la Inauguración Oficial de la Fábrica de Cemento 'Carlos Marx.'" May 29, Guaibaro, Cienfuegos.

———. 1980. "Discurso pronunciado . . . en la clausura del II Período Ordinario de Sesiones de la Asamblea Nacional del Poder Popular." December 27, Palacio de las Convenciones, Havana.

———. 1981. "Discurso pronunciado . . . en el Acto Central con Motivo del XXVIII Aniversario del Asalto al Cuartel Moncada." July 26, Las Tunas.

———. 1982. "Discurso pronunciado . . . en el Acto Central por el 29 Aniversario del Ataque al Cuartel Moncada." July 26, Bayamo.

———. 1983. "Discurso pronunciado . . . en la Inauguración del Combinado Textil de Santiago de Cuba 'Celia Sánchez Manduley.'" July 27, Santiago de Cuba.

———. 1984. "Discurso pronunciado . . . en el acto central por el XXXI Aniversario del Asalto al Cuartel Moncada." July 26, Cienfuegos.

———. 1984. "Discurso pronunciado . . . en el acto de Clausura del Primer Fórum Nacional de Energía." December 4, Teatro Carlos Marx, Havana.

———. 1986. "Discurso pronunciado . . . en la clausura de la Sesión diferida del Tercer Congreso del Partido Comunista de Cuba." December 2, Teatro Carlos Marx, Havana.

———. 1986. "Discurso pronunciado . . . en el Parque 'Céspedes', de Bayamo." December 19, Bayamo.

———. 1990. "Discurso pronunciado . . . en la clausura del V Congreso de La Federación de Mujeres Cubanas." March 7, Palacio de las Convenciones, Havana.

———. 1990. "Discurso pronunciado . . . en el Acto Central por el XXX Aniversario de los Comités de Defensa de la Revolución, efectuado en el Teatro 'Carlos Marx.'" September 28, Teatro Carlos Marx, Havana.

———. 1991. "Discurso pronunciado . . . en la Inauguración del IV Congreso del Partido Comunista de Cuba." October 10, Teatro Heredia, Santiago de Cuba.

———. 1992. "Discurso pronunciado . . . en el acto por el XXXIX Aniversario del Asalto al Cuartel Moncada y el XXXV del Levantamiento de Cienfuegos." September 5, Cienfuegos.

———. 1993. "Discurso pronunciado . . . en la clausura de la Sesión de Constitución de la Asamblea Nacional, en su cuarta legislatura, y del Consejo de Estado." March 15, Palacio de las Convenciones, Havana.

———. 1993. "Discurso pronunciado . . . en la clausura del acto central por el XL Aniversario del Asalto a los Cuarteles Moncada y 'Carlos Manuel de Céspedes.'" July 26, Teatro Heredia, Santiago de Cuba.

———. 2003. "Discurso del Comandante en Jefe . . . durante su visita al Grupo Empresarial Panda, Nanjing." February 28, Nanjing, China.

———. 2005. "Discurso pronunciado . . . en el acto por el aniversario 60 de su ingreso a la universidad." November 17, Universidad de La Habana, Havana.

———. 2006. "Discurso pronunciado . . . en ocasión del aniversario 47 de su entrada en Pinar del Río, el acto por la culminación del montaje de los grupos electrógenos en esa provincia." January 17, Pinar del Río.

———. 2006. "Discurso pronunciado . . . en el acto por el Día Internacional de los Trabajadores." May 1, Plaza de la Revolución, Havana.

———. 2006. "Discurso pronunciado . . . en la entrega de 101 vehículos a la Unión Eléctrica." May 5, Unión Eléctrica Nacional, Havana.

———. 2008 (1953). *La historia me absolverá (Edición anotada)*. Havana: Oficina de Publicaciones del Consejo de Estado.

Chávez Frías, Hugo. 2007. "Petrocaribe es parte del cambio de época que estamos viviendo." December 21, IV Summit of PetroCaribe, Cienfuegos. http://www.granma.cu/granmad/secciones/petrocaribe/de-la-iv-cumbre /art36.html.

Guevara, Ernesto Che. 1961. "Conferencia en el ciclo 'Economía y Planificación' de la Universidad Popular." April 30, Havana, José Altshuler private collection.

———. 1961. "Informe del Dr. Ernesto Che Guevara, Ministro de Industrias." *Obra Revolucionaria* 30: 107–28.

———. 1963. "Discurso en el Primer Fórum de Energía Eléctrica." November 22, Havana. https://www.youtube.com/watch?v=-z9hd49TkXY.

———. 1966 (1962) "Reuniones Bimestrales." In *El Che en la Revolución Cubana: Ministerio de Industrias* 6, edited by Ministerio del Azúcar, n.pp. Havana: Ministerio del Azúcar.

———. 1970 (1961). "Antonio Guiteras." In *Obras 1957–1967, II: La transformación política, económica y social*, 620–37. Havana: Casa de las Américas.

Lage Dávila, Carlos. 2007. "Puede decirse que en menos de tres años el país alcanzó una capacidad de generación eléctrica suficientemente por encima de su demanda máxima." *Granma*, 2nd ed., June 7: 3.

Lenin, Vladimir I. 1966 (1920). "Eighth All-Russia Congress of Soviets, 29 December." Translated by Julius Katzer. In *Collected Works* 31, 4th English ed., 461–534. Moscow: Progress Publishers.

Núñez Jiménez, Antonio. 1969. "El acto de inauguración del Instituto de Física Nuclear." *Academia de Ciencias de Cuba: Serie Física Nuclear* 1: 3–8.

NEWSPAPERS AND MAGAZINES

Administración Revolucionaria. 1960. "A todos los trabajadores eléctricos." *Guiteras*, September: 10–11.

Agencia Bolivariana de Noticias. 2010. "Chávez: Aportes de Cuba suman 10 veces más del costo del petróleo que envía Venezuela." *Agencia Bolivariana de Noticias*, April 17. http://www.aporrea.org/energia/n155408.html.

Agencia de Información Nacional. 1985. "Se registró un sobreconsumo de electricidad en los primeros veinte días de junio." *Granma*, 3rd ed., June 28: 1–2.

Alemany Gutiérrez, Edmundo. 2007. "Incumplidores y excedidos en planes de electricidad." *Guerrillero*, January 5: 2.

———. 2007. "La energía de una Revolución." *Guerrillero*, January 12: 8.

Antúnez, Rebeca. 1993. "Hacer en casa lo que antes costaba divisas." *Trabajadores*, January 11: 1.

———. 1993. "Los eléctricos no sólo dan apagones." *Trabajadores*, January 4: 3.

Bérriz Pérez, Luis. 2007. "Calentador solar de tubos al vacío." *Energía y tú* 39: 4–11.

Blanco, Marilyn. 1993. "Primer Taller Científico de Energía." *Guerrillero*, May 21: 3.

Bohemia. 2006. "Revolución energética en línea." *Bohemia*, May 12: 37.

Borrego, Juan Antonio. 1993. "Que la paja de arroz no vaya a los hornos de desperdicio." *Granma*, June 16: 2.

Camacho, Ledys. 1997. "Interviene Francia en modernización de centrales eléctricas." *Opciones*, November 6: 9.

Camacho, René. 1966. "Nunca antes un técnico tuvo tantas oportunidades como ahora.—Domenech." *Granma*, 3rd ed., July 3: 4.

Cardoso, Caima, and Ramón Brizuela. 2004. "Continúan las medidas alternativas." *Guerrillero*, August 16: 1.

Castro Morales, Yudy, and Ronald Suárez Rivas. 2019. "Otra mala noticia para el imperio: De la actual situación saldremos más fortalecidos." *Granma*, September 16. http://www.granma.cu/cuba/2019-09-16/como-atenuar-las -limitaciones-del-combustible-en-artemisa-analiza-el-presidente-cubano-16 -09-2019-13-09-01.

Castro Ruz, Fidel. 2005. "Medidas sobre las tarifas eléctricas, los incrementos salariales y los de la Seguridad y la Asistencia Social." *Juventud Rebelde*, 2nd ed., November 23: 1.

Coto Acosta, Lourdes. 1984. "Primer Fórum Nacional de Energía." *Energía* 1: 4–5.

Crespo Porbén, Ramón. 1993. "Valoran principales cuadros comportamiento del Programa Energético." *Guerrillero*, May 7: 8.

———. 1993. "Vivir de la candela." *Guerrillero*, June 4: 4.

Cuba Foreign Trade. 1998. "The Cuban Basic Industry: A Powerful Entrepreneurial System." *Cuba Foreign Trade* 2: 9–13.

de la Cruz, Oria. 1993. "'Moñito' ahorrativo." *Granma*, June 23: 3.

de la Torre, César. 1966. "El Partido en la construcción de la termoeléctrica de Mariel." *Granma*, June 28: 2.

del Barrio Menéndez, Emilio. 1973. "Unifican sistemas energéticos de occidente y oriente." *Granma*, 3rd ed., February 14: 4.

Energía. 1984. "Guía elaborada por la Comisión Nacional de Energía." *Energía* 1: 43–48.

———. 1985. "Declaración final del Primer Fórum Nacional de Energía." *Energía* 1–2: 13–18.

Enviados especiales de Novosti. 1963. "Bratsk: Poder soviético más electrificación igual al comunismo." *Rotograbado de Revolución* (supplement to *Revolución*), 2nd ed., June 3: 1–6.

García Elisalde, Alejandra. 2019. "La energía que renueva a Cuba." *Granma*, November 28. http://www.granma.cu/cuestion-de-leyes/2019-11-28/la -energia-que-renueva-a-cuba-28-11-2019-00-11-44.

Gómez, Marcelino. 1993. "Resolver los problemas de la población con el concurso de todos y concretar en cada sitio el Programa Energético." *Guerrillero*, June 18: 1, 8.

Granma. 1993. "Reseña de los debates." *Granma*, June 29: 2–4.

———. 1993. "Se está discutiendo el tema de la energía en medio de las condiciones más complejas." *Granma*, June 29: 5.

Guerrillero. 1993. "El Programa Energético Integral: Tarea a materializarse en cada colectivo laboral." *Guerrillero*, May 14: 8.

———. 1993. "Programación de apagones próxima semana." *Guerrillero*, May 28: 4.

———. 1993. "Informan sobre reajustes en la programación de apagones." *Guerrillero*, June 4: 7.

———. 2006. "Sobre la asignación del combustible doméstico para reserva de cada familia pinareña." *Guerrillero*, January 13: 2.

Gumá, José Gabriel. 1966. "100 mil kilovatios más para el consumo del país." *Granma*, 3rd ed., February 20: 4.

INRA. 1960. "¡Se llamaban!" *INRA* 1(8): 4–15.

Jiménez García, Eduardo. 2001. "Energía. El 2000 supo a petróleo." *Bohemia*, January 12: 22–26.

Juventud Rebelde. 2008. "El día después del paso de Gustav." *Juventud Rebelde*, September 2. http://www.juventudrebelde.cu/cuba/2008-09-02/el-dia -despues-delpaso-de-gustav.

———. 2008. "Restablecen paulatinamente servicio eléctrico en la Isla de la Juventud." *Juventud Rebelde*, September 4. http://www.juventudrebelde.cu /cuba/2008-09-04/restablecen-paulatinamenteservicio-electrico-en-la-isla -de-la-juventud.

———. 2019. "Cuba cuenta con un nuevo parque solar fotovoltaico." *Juventud Rebelde*, June 25. http://www.juventudrebelde.cu/ciencia-tecnica/2019-06-25 /cuba-cuenta-con-un-nuevo-parque-solar-fotovoltaico.

Lapaz, Victor, and Luis Bérriz. 1997. "Energía solar: El camino de la vida (Entrevista a presidente de CubaSolar, Doctor Luis Bérriz)." *Energía y tú* 0, digital edition, n.pp.

Lee, Susana. 1993. "El vital tema de la energía, desde la base." *Granma*, June 19: 5.

León Moya, Haydée. 2004. "Continúa el trabajo para restablecer la estabilidad." *Granma*, September 2: 3.

López, Flor M. 1993. "Convocan a concurso sobre ahorro de energía." *Guerrillero*, May 21: 7.

López, Flor M., and Luis Manuel Avila. 1993. "Acuerdan asambleas municipales del Poder Popular medidas para concretar ahorro de energético y perfeccionar los consejos." *Guerrillero*, June 11: 7.

Martínez García, Yisel. 2019. "Combustible, problema sin resolver." *Granma*, April 12. http://www.granma.cu/cuba/2019-04-12/combustible-problema-sin -resolver-12-04-2019-20-04-50.

Mayoral, María Julia. 2004. "Detallada información al pueblo sobre los problemas en el servicio eléctrico." *Granma*, September 28: 1, 8.

———. 2007. "Revolución Energética con resultados loables." *Granma*, 2nd ed., December 20: 3.

Mayoral, María Julia, and José A. de la Osa. 2004. "La población ha sido, es y será lo más sagrado para la Revolución." *Granma*, September 29: 1, 8.

Molina Pérez, Yolanda. 2004. "Reconoce Lage esfuerzo de los pinareños." *Guerrillero*, August 17: 1.

———. 2004. "De luz y de sombra." *Guerrillero*, August 21: 4.

Monduy Cinteo, Alberto. 1993. "Analizan situación de los combustibles y Programa Energético." *Guerrillero*, June 4: 8.

Monteagudo, Katia. 2006. "Paso Real no cree en apagones." *Bohemia*, May 12: 19–21.

Monteagudo, Katia, and Ariel Terrero. 2007. "En Cuba Revolución Energética: La luz que en tus ojos busco." *Bohemia*, June 22: 28–36.

Nicado, Clemente. 1993. "¿Cómo 'estirar' el keroseno?" *Granma*, May 29: 3.

Oramas, Joaquín. 1980. "Entrará en vigor el primero de enero la nueva tarifa eléctrica para centros industriales, de servicios, administrativos y privados." *Granma*, 3rd ed., November 25: 2.

———. 1980. "La nueva tarifa eléctrica busca el ahorro de petróleo y eliminar el despilfarro de electricidad." *Granma*, 3rd ed., March 18: 4–5.

———. 1980. "La nueva tarifa eléctrica entra en vigor en octubre." *Granma*, 3rd ed., September 30: 1, 4.

———. 1980. "Podrá significar inicialmente el ahorro de 700 millones de KWh— más de 200 mil toneladas de petróleo la aplicación de la nueva tarifa eléctrica." *Granma*, 3rd ed., March 29: 3.

———. 1985. "Disminuyeron los consumos energéticos en La Habana en el primer trimestre." *Granma*, 3rd ed., April 30: 1.

———. 1985. "Disminuyó en enero el consumo de electricidad en Ciudad de La Habana en 20 541,4 megawatts-hora." *Granma*, 3rd ed., March 2: 2.

———. 1986. "Se extiende ya de punta a punta de Cuba el sistema electroenergético nacional." *Granma*, 1st ed., June 23: 1.

Pérez, Dora, Agnerys Rodríguez, Julieta García, Alina Perera, Norges Martínez, Amaury E. Valle, and Ricardo Ronquillo. 2005. "La Revolución da un paso hacia su invulnerabilidad política." *Juventud Rebelde*, November 24: 1, 4.

Prijodko, B., and M. Marcer. 1961. "La industria electroenergética soviética." *INRA* 2(6): 60–65.

Regalado, Zenia. 2004. "Se activan pozos con bombas de mano." *Guerrillero*, August 19: 1.

———. 2004. "Permanecerán activas plantas eléctricas." *Guerrillero*, August 22: 2.

Regalado, Zenia, and Edmundo Alemany. 2004. "La electricidad se restablece." *Guerrillero*, August 24: 1.

Revolución. 1959. "Lo mejor para mamá." *Revolución*, May 8: 6.

———. 1959. "Mayo 10: Día de las Madres." *Revolución*, May 8: 23.

———. 1959. "Rebajan tarifas eléctricas." *Revolución*, 2nd ed., August 20: 1, 19.

———. 1960. "Su preferencia por este monograma está plenamente justificada. . . ." *Revolución*, pt. 3, January 1: 16.

———. 1960. "La Cía. Electric de Cuba se enorgullece en brindarle cocinas y calentadores cubanos Wesco." *Revolución*, April 18: 7.

———. 1960. "24 millones ahorrará Cuba en compra de petróleo." *Revolución*, 2nd ed., May 27: 1.

———. 1960. "Agresión política a Cuba la actitud del 'trust petrolero.'" *Revolución*, 2nd ed., June 13: 14.

———. 1960. "Cofiño, Fraginals y la Embajada yanqui." *Revolución*, 1st ed., December 13: 1.

———. 1960. "Expulsa el 26-7 a Amaury Fraginals y a F. Iglesias." *Revolución*, 2nd ed., December 12: 1.

———. 1960. "La primera gran zancadilla contra nuestra Revolución." *Revolución*, 2nd ed., June 11: 1, 8, 20.

———. 1960. "Orden a la Texaco de procesar petróleo del Estado." *Revolución*, 2nd ed., June 29: 1.

———. 1960. "Orden de refinar a la Esso y la Shell." *Revolución*, 2nd ed., July 1: 1, 6.

———. 1960. "Ratificarán esta noche los obreros eléctricos su apoyo a la Revolución." *Revolución*, 1st ed., December 14: 1.

———. 1960. "Van a saber lo que es una Revolución. A sacudir la mata con mano firme—dice Fidel." *Revolución*, 1st ed., December 15: 1.

———. 1961. "Cuba revolucionaria ganará también la batalla de los abastecimientos. Esfuerzo común para la mayor producción." *Revolución*, 1st ed., July 6: 2–4, 10.

Rodríguez Martínez, Osvaldo. 1993. "La energía en Cuba: Noticias reales." *Juventud Rebelde*, June 27: 6–7.

Sánchez, Lorena, and Oscar Sánchez Serra. 2014. "Fuentes renovables de energía: Abre camino de la actualización." *Granma*, November 7: 4–5.

Sánchez Serra, Oscar. 2021. "La agricultura urbana, suburbana y familiar: La comida más cerca de casa." *Granma*, 24 April. https://www.granma.cu/cuba/2021-04-24/la-agricultura-urbana-suburbana-y-familiar-la-comida-mas-cerca-de-casa-24-04-2021-01-04-58.

Sartorio, Blanchie. 2006. "El amor entró en la cocina." *Guerrillero*, January 13: 8.

Suárez, Ronal. 2004. "Batalla de alta tensión." *Guerrillero*, August 18: 4.

———. 2004. "Agua desde la refinería del 'Martí.'" *Guerrillero*, August 19: 4.

Suárez Rivas, Ronald. 2007. "Kilowatts en saco roto." *Granma*, 2nd ed., December 7: 8.

———. 2009. "Desmesurado gasto eléctrico por los organismos en Pinar del Río." *Granma*, May 20: 3.

Tamayo León, René, and Edmundo Alemany Gutiérrez. 2004. "Tenemos los recursos para recuperarnos." *Guerrillero*, August 15: 1.

Terrero, Ariel. 1992. "¿Cuántas patas tiene el gato?" *Bohemia*, August 14: 28–32.

Villaescusa, Ivette. 1993. "Asambleas sindicales para ahorrar energía." *Granma*, May 13: 1.

Werniuk, Jane. 2008. "Marching to a Different Drum." *Canadian Mining Journal*, February 1. http://www.canadianminingjournal.com/features/marching-to-a-differentdrum/1000220856.

OTHER PRIMARY SOURCES

Altshuler, José, Mario Alberto Arrastía Ávila, Ricardo Bérriz Valle, Ramiro Guerra Valdés, Víctor Bruno Henríquez Pérez, Luis Manuel Hernández García, Conrado Moreno Figueredo, Diosdado Pérez Franco, Víctor Omar Puentes Montó, Julio Torres Martínez, and Elena Vigil Santos. 2004. *Suplemento especial. [Tabloide para el curso de energía de la Universidad para Todos]*. Havana: Editorial Academia.

Anonymous. 1954. "Lo que fué y lo que es la . . . Cía, Cubana de Electricidad." In *Libro de Cuba: Una enciclopedia ilustrada que abarca las Artes, las Letras, las Ciencias, la Economía, la Política, la Historia, la Docencia y el Progreso General de la Nación Cubana*, edited by Juan Joaquín Otero, Juan Sabates, Rufino Pazos, Arturo Alfonso Roselló, and Bienvenido Madán y Estrada, 819–21. Havana: Talleres litográficos de Artes Gráficas.

Arrastía Ávila, Mario Alberto. 2006. "Preguntas y respuestas sobre el ahorro de energía eléctrica." In *Educación científica y energética: Importancia para la Revolución Energética en Cuba. IV Congreso Internacional de Didáctica de las Ciencias*, edited by Mario Alberto Arrastía Ávila, Roberto González Vale, Juan Fundora Lliteras, and Nidia Justa Mainegra Naranjo, 79–86. Havana: CubaSolar.

Ávila, Tony. 2012. "Regalao murió en el 80." In *Rodilla en tierra* (music album). Havana: BisMusic.

Bérriz Pérez, Luis. 2006. "Consideraciones sobre el desarrollo histórico del uso de las fuentes renovables de energía, a partir del triunfo de la Revolución Cubana." In *Educación científica y energética: Importancia para la Revolución Energética en Cuba. IV Congreso Internacional de Didáctica de las Ciencias*, edited by Mario Alberto Arrastía Ávila, Luis Bérriz Pérez, Roberto González Vale, Juan Fundora Lliteras, and Nidia Justa Mainegra Naranjo, 46–54. Havana: CubaSolar.

Bérriz, Luis, and Emir Madruga. 2000. *Cuba y las fuentes renovables de energía*, 4th ed. Havana: Editora de CubaSolar.

Castro Díaz-Balart, Fidel. 1990. "La energía nuclear en Cuba: Factor imprescindible para el desarrollo." *Boletín del OIEA* 1: 49–52.

———. 2005. *Nuclear Energy: Environmental Danger or Solution for the 21st Century?* Turin, Italy: A. M. Arti Grafiche.

DURE. n.d. *La energía y mi país: Una Revolución con energía.* Havana: Dirección de Uso Racional de la Energía de la Unión Eléctrica (MINBAS).

EIE. n.d. "ATEC Systems: Nuestra Empresa." Empresa de la Industria Electrónica, http://www.eie.co.cu/empresa.html.

FANJ. 1966–67. *Energía nuclear 1966, 1967* (letter collection). Havana: Fundación "Antonio Núñez Jiménez" de la Naturaleza y el Hombre.

Gaiga, Padre Joaquín. 2008. *La cruz al pie de los mogotes: Apuntes para la historia de Viñales.* Pinar del Río: Ediciones Vitral.

González Alonso, Edgardo F., and Carlos March Alvarez-Muñoz. 1982. "Visión perspectiva del suministro energético nacional." *Academia de Ciencias de Cuba: Boletín de Energía Solar* 3: 1–33.

González Jordán, Roberto. 1986. *Ahorro de Energía en Cuba.* Havana: Editorial Científico-Técnica.

Guevara, Ernesto Che. 2003 (1961). "Cuba: Historical Exception or Vanguard in the Anticolonial Struggle?" In *The Che Guevara Reader: Writings on Politics and Revolution*, 2nd ed., edited by David Deutschmann, 130–42. Melbourne, Australia: Ocean Press.

Guzón Camporredondo, Ada, ed. 2006. *Desarrollo local en Cuba: Retos y perspectivas.* Havana: Editorial Academia.

IBRD. 1951. *Report on Cuba.* Washington, DC: International Bank for Reconstruction and Development.

MINBAS. 2001. *Ahorro de energía: La esperanza del futuro. Para maestros: Primer ciclo de la Educación Primaria y Especial.* Havana: Editora Política.

———. 2002. *Ahorro de energía y respeto ambiental: Bases para un futuro sostenible. Libro del Programa de Ahorro de Electricidad en Cuba para la enseñanza media.* Havana: Editora Política.

Moreno Figueredo, Conrado. 2011. *Energía eólica: Tecnología y aplicaciones.* Havana: Editorial Academia.

Núñez Jiménez, Antonio. 1959. *Geografía de Cuba: Adaptada al Nuevo Programa Revolucionario de Bachillerato*, 2nd ed. Havana: Editorial Lex.

Pagés Díaz, Jhosané. 2010. "Potencialidad de residuos agroindustriales para producir biogás: Caso de estudio EPG 'Camilo Cienfuegos.'" MSc diss., Instituto Superior Politécnico "José Antonio Echeverría" (ISPJAE), Cuba.

PCC. 1975. *I Congreso del PCC: Tesis y Resoluciones sobre las directivas para el desarrollo económico y social en el quinquenio 1976–1980*. Havana: Partido Comunista de Cuba.

———. 1980. *II Congreso del Partido Comunista de Cuba (Informe central)*. Havana: Partido Comunista de Cuba.

Pedrera, Ramón. 2008. "Gestión energética en Empresa Eléctrica de Pinar del Río. Empresa Eléctrica Provincial, Calle Máximo Gómez #38, Pinar del Río." MSc diss., Universidad de Pinar del Río "Hermanos Saíz Montes de Oca," Cuba.

PetroCaribe. 2005. "Acuerdo de Cooperación Energética Petrocaribe." June 29, Puerto La Cruz, Venezuela. http://www.granma.cu/granmad/secciones/petrocaribe/cumbres-cel/acuerdo-1.html.

———. 2014. *PetroCaribe Management Report Quarter I*. Caracas: PDV Caribe.

Pons Duarte, Hugo M. 1988. *Política energética, política económica y desarrollo*. Havana: Editora Política.

Presa, Juan. 2008. *La Revolución Energética: Resultados y perspectivas. Enero 2008*. Powerpoint presentation, UNE Director of Generation, Feria Internacional del Libro, Havana.

Sánchez Almeira, Yosvany. 2006. "Gestión energética empresarial: Empresa Azucarera Harlem, Bahía Honda." MSc diss., Universidad de Pinar del Río "Hermanos Saíz Montes de Oca," Cuba.

Semevskiy, Boris N. 1967. "The Problem of Cuba's Energy Supplies." *Soviet Geography* 8(1): 25–33.

Suárez Rodríguez, José Antonio, and Pedro Aníbal Beatón Soler. 2007. "Estado y perspectivas de las energías renovables en Cuba." *Tecnología Química* 27(3): 75–82.

Torres-Martínez, Julio. 2006 (2005). "Nuevo Sistema Electroenergético Nacional en Cuba, basado fundamentalmente en biomasa cañera." In *El camino del sol: Un desafío para la humanidad en el tercer milenio. Una esperanza para los países del sur*, edited by Enrico Turrini, 322–55. Havana: Editorial CubaSolar.

Turrini, Enrico. 2006. *El camino del sol: Un desafío para la humanidad en el tercer milenio. Una esperanza para los países del sur*. Havana: Editorial CubaSolar.

Viego Felipe, Percy, Marcos de Armas Teyra, Ignacio Pérez Abril, Arturo Padrón Padrón, and Leonardo Casas Fernández. 2007. *Temas especiales de sistemas eléctricos industriales*. Cienfuegos: Editorial Universo Sur.

Wettstein, Germán. 1969. *Vivir en Revolución: 20 semanas en Cuba*. Montevideo, Uruguay: Editorial Signo.

STATISTICAL COLLECTIONS

BP. 2022. *Statistical Review of World Energy: All Data, 1965–2021*. https://www
.bp.com/content/dam/bp/business-sites/en/global/corporate/xlsx/energy
-economics/statistical-review/bp-stats-review-2022-all-data.xlsx.
CMEA Secretariat. 1979. *Statistical Yearbook of Member States of the Council
for Mutual Economic Assistance 1979*. London: IPC Industrial Press.
Comité Estatal de Estadísticas. 1989. *Anuario Estadístico de Cuba de 1989*.
Havana: Editorial Estadística.
EIA. 2015. "Cuba. Analysis—Energy Sector Highlights, January 2015." http://
www.eia.gov/beta/international/country.cfm?iso=CUB.
IEA. 2018. "Country Data: Cuba." https://www.iea.org/statistics/?country=CUBA.
MPetroMin. 2013–14. *Petróleo y otros estadísticos*. Caracas: Ministerio del
Poder Popular de Petróleo y Minería.
ONE. 2009. *Estadísticas energéticas en la Revolución*. Havana: Oficina
Nacional de Estadísticas.
ONEI. 2014–20. *Anuario estadístico de Cuba*. Havana: Oficina Nacional de
Estadística e Información.
OPEC. 2016. *Annual Statistical Bulletin 2016*. Vienna: Organization of the
Petroleum Exporting Countries.

LITERATURE

Adams, Richard N. 1975. *Energy and Structure: A Theory of Social Power*.
Austin: University of Texas Press.
Aklin, Michaël, and Johannes Urpelainen. 2018. *Renewables: The Politics of a
Global Energy Transition*. Cambridge, MA: MIT Press.
Alexander, Robert J. 2002. *A History of Organized Labor in Cuba*. Westport,
CT: Praeger.
Alonso-Pippo, Walfrido, Carlos A. Luengo, John Koehlinger, Pietro Garzone,
and Giacinto Cornacchia. 2008. "Sugarcane Energy Use: The Case of Cuba."
Energy Policy 36: 2163–81.
Altieri, Miguel A. 1995. *Agroecology: The Science of Sustainable Agriculture*,
2nd ed. Boulder, CO: Westview Press.
Altieri, Miguel A., Nelso Companioni, Kristina Cañizares, Catherine Murphy,
Peter Rosset, Martin Bourque, and Clara I. Nicholls. 1999. "The Greening of
the 'Barrios': Urban Agriculture for Food Security in Cuba." *Agriculture and
Human Values* 16(2): 131–40.
Altshuler, José. 2014. *Las comunicaciones internacionales de Cuba: Del correo
al satélite*. Havana: Editorial Científico-Técnica.
Altshuler, José, and Miguel González. 1997. *Una luz que llegó para quedarse:
Comienzos del alumbrado eléctrico y su introducción en Cuba*. Havana:

Editorial Científico-Técnica and Oficina del Historiador de la Ciudad de La Habana.

Alvarez, José. 2004. *Cuba's Agricultural Sector*. Gainesville: University of Florida Press.

Anand, Nikhil, Akhil Gupta, and Hannah Appel, eds. 2018. *The Promise of Infrastructure*. Durham, NC: Duke University Press.

Anusas, Mike, and Tim Ingold. 2015. "The Charge against Electricity. *Cultural Anthropology* 30(4): 540–54.

Aronoff, Kate, Alyssa Battistoni, Daniel Aldana Cohen, and Thea Riofrancos. 2019. *A Planet to Win: Why We Need a Green New Deal*. London: Verso.

Arrastía-Ávila, Mario Alberto, and Lisa M. Glidden. 2017. "Cuba's Energy Revolution and 2030 Policy Goals: More Penetration of Renewable Energy in Electricity Generation." *International Journal of Cuban Studies* 9(1): 73–90.

Azicri, Max. 2009. "The Castro-Chávez Alliance." *Latin American Perspectives* 36(1): 99–110.

Bain, Mervyn J. 2001. "Soviet/Cuba Relations 1985–1991." PhD diss., University of Glasgow, United Kingdom.

Baka, Jennifer, and Saumya Vaishnava. 2020. "The Evolving Borderland of Energy Geographies." *Geography Compass* 14: 1–17.

Baker, Lucy, Peter Newell, and Jon Phillips. 2014. "The Political Economy of Energy Transitions: The Case of South Africa." *New Political Economy* 19(6): 791–818.

Banerjee, Anindita. 2003. "Electricity: Science Fiction and Modernity in Early Twentieth-Century Russia." *Science Fiction Studies* 30(1): 49–71.

Barak, On. 2020. *Powering Empire: How Coal Made the Middle East and Sparked Global Carbonization*. Oakland: University of California Press.

Barca, Stefania. 2019. "The Labor(s) of Degrowth". *Capitalism Nature Socialism* 30(2): 207–16.

Barkin, David. 1980. "Confronting the Separation of Town and Country in Cuba." *Antipode* 12(3): 31–40.

Barnes, Douglas F., and Willem M. Floor. 1996. "Rural Energy in Developing Countries: A Challenge for Economic Development." *Annual Review of Energy and the Environment* 21: 497–530.

Barnes, Jessica. 2017. "States of Maintenance: Power, Politics, and Egypt's Irrigation Infrastructure." *Environment and Planning D: Society and Space* 35(1): 146–64.

Barry, Andrew. 2015. "Thermodynamics, Matter, Politics." *Distinktion: Journal of Social Theory* 16(1): 110–25.

Bell, Karen. 2011. "Environmental Justice in Cuba." *Critical Social Policy* 31(2): 241–65.

Bengelsdorf, Carollee. 1994. *The Problem of Democracy in Cuba: Between Vision and Reality*. Oxford: Oxford University Press.

Benjamin-Alvarado, Jonathan. 2000. *Power to the People: Energy and the Cuban Nuclear Program*. New York: Routledge.

Bimber, Bruce. 1994. "Three Faces of Technological Determinism." In *Does Technology Drive History? The Dilemma of Technological Determinism*, edited by Merritt Roe Smith and Leo Marx, 79–100. Cambridge, MA: MIT Press.

Black, Brian. 2000. *Petrolia: The Landscape of America's First Oil Boom*. Baltimore, MD: Johns Hopkins University Press.

Blaikie, Piers M., and Harold Brookfield, eds. 1987. *Land Degradation and Society*. London: Methuen.

Blondeel, Mathieu, Michael J. Bradshaw, Gavin Bridge, and Caroline Kuzemko. 2021. "The Geopolitics of Energy System Transformation: A Review." *Geography Compass* 15: e12580.

Blue, Sarah A. 2007. "The Erosion of Racial Equality in the Context of Cuba's Dual Economy." *Latin American Politics and Society* 44(3): 35–68.

Bobes, Velia Cecilia. 2016. "Reformas en Cuba: ¿Actualización del socialismo o reconfiguración social?" *Cuban Studies* 44: 165–88.

Bogomólov, Oleg. 1986. *La industria energética sin crisis: Por qué no ha habido crisis energética en los países del CAME*. Moscow: Editorial de la Agencia de Prensa Nóvosti.

Boillat, Sébastien, Julien-François Gerber, and Fernando R. Funes-Monzote. 2012. "What Economic Democracy for Degrowth? Some Comments on the Contribution of Socialist Models and Cuban Agroecology." *Futures* 44(6): 600–607.

Borowy, Iris. 2013. "Degrowth and Public Health in Cuba: Lessons from the Past?" *Journal of Cleaner Production* 38: 17–26.

Bouzarovski, Stefan, Sergio Tirado Herrero, Saska Petrova, Jan Frankowski, Roman Matoušek, and Tomas Maltby. 2017. "Multiple Transformations: Theorizing Energy Vulnerability as a Socio-Spatial Phenomenon." *Geografiska Annaler: Series B, Human Geography*: 99(1): 20–41.

Boyer, Dominic. 2014. "Energopower: An Introduction." *Anthropological Quarterly* 87(2): 309–33.

———. 2018. "Infrastructure, Potential Energy, Revolution." In *The Promise of Infrastructure*, edited by Nikhil Anand, Akhil Gupta, and Hannah Appel, 221–43. Durham, NC: Duke University Press.

Bridge, Gavin. 2018. "The Map Is Not the Territory: A Sympathetic Critique of Energy Research's Spatial Turn." *Energy Research and Social Science* 36: 11–20.

Bridge, Gavin, Stefan Bouzarovski, Michael Bradshaw, and Nick Eyre. 2013. "Geographies of Energy Transition: Space, Place and the Low-Carbon Economy." *Energy Policy* 53: 331–40.

Bridge, Gavin, Begüm Özkaynak, and Ethemcan Turhan. 2018. "Energy Infrastructure and the Fate of the Nation: Introduction to the Special Issue." *Energy Research and Social Science* 41: 1–11.

Bridge, Gavin, and Andrew Wood. 2010. "Less Is More: Spectres of Scarcity and the Politics of Resource Access in the Upstream Oil Sector." *Geoforum* 41(4): 565–76.

Bryant, Raymond L., and Sinéad Bailey. 1997. *Third World Political Ecology.* London: Routledge.

Bunker, Stephen G. 1985. *Underdeveloping the Amazon: Extraction, Unequal Exchange, and the Failure of the Modern State.* Urbana: University of Illinois Press.

Bustamante, Michael J., and Jennifer L. Lambe, eds. 2019. *The Revolution from Within: Cuba, 1959–1980.* Durham, NC: Duke University Press.

Cabrera Arús, María A. 2019. "The Material Promise of Socialist Modernity: Fashion and Domestic Space in the 1970s." In *The Revolution from Within*, edited by Michael J. Bustamante and Jennifer L. Lambe, 189–217. Durham, NC: Duke University Press.

Calder, Kent E. 2012. *The New Continentalism: Energy and Twenty-First Century Eurasian Geopolitics.* New Haven, CT: Yale University Press.

Carse, Ashley. 2014. *Beyond the Big Ditch: Politics, Ecology, and Infrastructure at the Panama Canal.* Cambridge, MA: MIT Press.

Castán Broto, Vanesa. 2019. *Urban Energy Landscapes.* Cambridge: Cambridge University Press.

Castán Broto, Vanesa, Idalina Baptista, Joshua Kirshner, Shaun Smith, and Susana Neves Alves. 2018. "Energy Justice and Sustainability Transitions in Mozambique." *Applied Energy* 228: 645–55.

Cederlöf, Gustav. 2016. "Low-Carbon Food Supply: The Ecological Geography of Cuban Urban Agriculture and Agroecological Theory." *Agriculture and Human Values* 33(4): 771–84.

———. 2020. "Maintaining Power: Decarbonisation and Recentralisation in Cuba's Energy Revolution." *Transactions of the Institute of British Geographers* 45(1): 81–94.

———. 2020. "The Revolutionary City: Socialist Urbanisation and Nuclear Modernity in Cienfuegos, Cuba." *Journal of Latin American Studies* 52(1): 53–76.

———. 2021. "Out of Steam: Energy, Materiality, and Political Ecology." *Progress in Human Geography* 45(1): 70–87.

Cederlöf, Gustav, and Alf Hornborg. 2021. "System Boundaries as Epistemological and Ethnographic Problems: Assessing Energy Technology and Socio-Environmental Impact." *Journal of Political Ecology* 28: 111–23.

Cederlöf, Gustav, and Donald V. Kingsbury. 2019. "On PetroCaribe: Petro-politics, Energopower, and Post-Neoliberal Development in the Caribbean Energy Region." *Political Geography* 72: 124–33.

Cheng, Yinghong. 2007. "Fidel Castro and 'China's Lesson for Cuba': A Chinese Perspective." *The China Quarterly* 189: 24–42.

Cherni, Judith A., and Yohan Hill. 2009. "Energy and Policy Providing for Sustainable Livelihoods in Remote Locations—the Case of Cuba." *Geoforum* 40(4): 645–54.

Clark, Eric. 2009. "Island Development." In *International Encyclopedia of Human Geography* 5, edited by Rob Kitchin and Nigel Thrift, 607–10. Amsterdam: Elsevier.

Clark, Eric, and Thomas L. Clark. 2009. "Isolating Connections—Connecting Isolations." *Geografiska Annaler: Series B, Human Geography* 91(4): 311–23.

Cocq, Karen, and David A. McDonald. 2010. "Minding the Undertow: Assessing Water 'Privatization' in Cuba." *Antipode* 42(1): 6–45.

Cooper, R. Caron. 1986. "Petroleum Displacement in the Soviet Economy: The Case of Electric Power Plants." *Soviet Geography* 27(6): 377–97.

Coopersmith, Jonathan. 1992. *The Electrification of Russia, 1880–1926*. Ithaca, NY: Cornell University Press.

Córdova, Efrén. 1987. *Castro and the Cuban Labour Movement: Statecraft and Society in a Revolutionary Period (1959–1961)*. Lanham, MD: University Press of America.

Coronil, Fernando. 1997. *The Magical State: Nature, Money, and Modernity in Venezuela*. Chicago: University of Chicago Press.

Crosby, Alfred W. 2004. *Ecological Imperialism: The Biological Expansion of Europe, 900–1900*, 2nd ed. Cambridge: Cambridge University Press.

———. 2006. *Children of the Sun: A History of Humanity's Unappeasable Appetite for Energy*. New York: W. W. Norton.

Daggett, Cara New. 2019. *The Birth of Energy: Fossil Fuels, Thermodynamics, and the Politics of Work*. Durham, NC: Duke University Press.

Dalakoglou, Dimitris. 2017. *The Road: An Ethnography of (Im)Mobility, Space, and Cross-Border Infrastructures in the Balkans*. Manchester: Manchester University Press.

D'Alisa, Giacomo, Federico Demaria, and Giorgos Kallis, eds. 2014. *Degrowth: A Vocabulary for a New Era*. London: Routledge.

Deere, Diana Carmen. 1993. "Cuba's National Food Program and Its Prospects for Food Security." *Agriculture and Human Values* 10(3): 35–51.

de la Fuente, Alejandro. 2001. "Recreating Racism: Race and Discrimination in Cuba's 'Special Period.'" *Socialism and Democracy* 15(1): 65–91.

Demaria, Federico, and Ashish Kothari. 2017. "The Post-Development Dictionary Agenda: Paths to the Pluriverse." *Third World Quarterly* 38(12): 2588–99.

Dove, Michael R., and Daniel M. Kammen. 2015. *Science, Society and the Environment: Applying Anthropology and Physics to Sustainability*. London: Routledge.

Dukes, Jeffrey S. 2003. "Burning Buried Sunshine: Human Consumption of Ancient Solar Energy." *Climatic Change* 61: 31–44.

Eckstein, Susan. 1990. "The Rectification of Errors or the Errors of the Rectification Process in Cuba?" *Cuban Studies* 20: 67–85.

———. 2003. *Back from the Future: Cuba under Castro*, 2nd ed. New York: Routledge.

Edgerton, David. 2006. *The Shock of the Old: Technology and Global History since 1900*. Oxford: Oxford University Press.

Emmanuel, Arghiri. 1972. *Unequal Exchange: A Study of the Imperialism of Trade*. Translated by Brian Pearce. New York: Monthly Review Press.

Engels, Friedrich. 1978 (1872). "On Authority." In *The Marx-Engels Reader*, 2nd ed., edited by Robert C. Tucker, 730–33. New York: W. W. Norton.

Enoch, Marcus, James P. Warren, Humberto Valdes Rios, and Enrique Henríquez Menoyo. 2004. "The Effect of Economic Restrictions on Transport Practices in Cuba." *Transport Policy* 11(1): 67–76.

Enríquez, Laura J. 2003. "Economic Reform and Repeasantization in Post-1990 Cuba." *Latin American Research Review* 38(1): 202–18.

Faris, Ali S., Saadi K. Al-Naseri, Nather Jamal, Raed Isse, Mezher Abed, Zainab Fouad, Akeel Kazim, Nihad Reheem, Ali Chaloob, Hazim Mohammad, Hayder Jasim, Jaafar Sadeq, Ali Salim, and Aws Abas. 2012. "Effects of Magnetic Field on Fuel Consumption and Exhaust Emissions in Two-Stroke Engines." *Energy Procedia* 18: 327–38.

Feinberg, Richard. 2013. "Foreign Investment in the New Cuban Economy." *NACLA: Report on the Americas* 46(1): 13–18.

Fischhendler, Itay, Lior Herman, and Lioz David. 2022. "Light at the End of the Panel: The Gaza Strip and the Interplay between Geopolitical Conflict and Renewable Energy Transition." *New Political Economy* 27(1): 1–18.

Fitzpatrick, Sheila. 2008. *The Russian Revolution*, 3rd ed. Oxford: Oxford University Press.

Fletcher, Tom. 2017. "La Revolución Energética: A Model for Reducing Cuba's Dependence on Venezuelan Oil." *International Journal of Cuban Studies* 9(1): 91–116.

Folch, Christine. 2013. "Surveillance and State Violence in Stroessner's Paraguay: Itaipú Hydroelectric Dam Archive of Terror." *American Anthropologist* 115(1): 44–57.

Font, Mauricio A. 2008. "Cuba and Castro: Beyond the 'Battle of Ideas.'" In *A Changing Cuba in a Changing World*, edited by Mauricio A. Font, 43–72. New York: Bildner Publication.

Forrest, David Peter. 1999. "*Bichos, Maricones* and *Pingueros*: An Ethnographic Study of Maleness and Scarcity in Contemporary Socialist Cuba." PhD diss., SOAS, University of London, United Kingdom.

Friedrichs, Jörg. 2010. "Global Energy Crunch: How Different Parts of the World Would React to a Peak Oil Scenario." *Energy Policy* 38(8): 4562–69.

Funes, Fernando, Luis García, Martin Bourque, Nilda Pérez, and Peter Rosset, eds. 2002. *Sustainable Agriculture and Resistance: Transforming Food Production in Cuba*. Oakland, CA: Food First Books.

Funes Monzote, Reinaldo. 2008. *From Rainforest to Cane Field in Cuba: An Environmental History since 1492*. Translated by Alex Martin. Chapel Hill: University of North Carolina Press.

———. 2019. "*Geotransformación*: Geography and Revolution in Cuba from the 1950s to the 1960s." In *The Revolution from Within: Cuba 1959–1980*, edited by Michael J. Bustamante and Jennifer L. Lambe, 117–45. Durham, NC: Duke University Press.

Galeano, Eduardo. 1997 (1973). *Open Veins of Latin America: Five Centuries of the Pillage of a Continent*. Translated by Cedric Belfrage. New York: Monthly Review Press.

García, Guadalupe. 2015. *Beyond the Walled City: Colonial Exclusion in Havana*. Oakland: University of California Press.

Gebelein, Jennifer. 2012. *A Geographic Perspective of Cuban Landscapes*. Dordrecht, Netherlands: Springer.

Georgescu-Roegen, Nicholas. 1971. *The Entropy Law and the Economic Process*. Cambridge, MA: Harvard University Press.

Gerber, Julien-François. 2020. "Degrowth and Critical Agrarian Studies." *Journal of Peasant Studies* 47(2): 235–64.

Gettig, Eric T. 2016. "Oil and Revolution in Cuba: Development, Nationalism, and the U.S. Energy Empire, 1902–1961." PhD diss., Georgetown University, United States.

Gleijeses, Piero. 2013. *Visions of Freedom: Havana, Washington, Pretoria, and the Struggle for Southern Africa, 1976–1991*. Chapel Hill: University of North Carolina Press.

Gómez-Baggethun, Erik. 2020. "More Is More: Scaling Political Ecology within Limits to Growth." *Political Geography* 76: 102095.

Gordy, Katherine. 2006. "'Sales + Economy + Efficiency = Revolution'? Dollar-ization, Consumer Capitalism, and Popular Responses in Special Period Cuba." *Public Culture* 18(2): 383–412.

Gudynas, Eduardo. 2016. "Beyond Varieties of Development: Disputes and Alternatives." *Third World Quarterly* 37(4): 721–32.

Guerra, Lillian. 2012. *Visions of Power in Cuba: Revolution, Redemption, and Resistance, 1959–1971*. Chapel Hill: University of North Carolina Press.

Gupta, Akhil. 2019. "The Future in Ruins: Thoughts on the Temporality of Infrastructure." In *The Promise of Infrastructure*, edited by Nikhil Anand, Akhil Gupta, and Hannah Appel, 62–79. Durham, NC: Duke University Press.

Gustafson, Thane. 1989. *Crisis amid Plenty: The Politics of Soviet Energy under Brezhnev and Gorbachev*. Princeton, NJ: Princeton University Press.

Hansing, Katrin. 2011. "Changes from Below: New Dynamics, Spaces, and Attitudes in Cuban Society." *NACLA: Report on the Americas* 44(4): 16–19.

Harris, Leila M. 2012. "State as Socionatural Effect: Variable and Emergent Geographies of the State in Southeastern Turkey." *Comparative Studies of South Asia, Africa and the Middle East* 32(1): 25–39.

Harrison, Conor. 2013. "The Historical-Geographical Construction of Power: Electricity in Eastern North Carolina." *Local Environment* 18(4): 469–86.

Harrison, Conor, and Jeff Popke. 2018. "Geographies of Renewable Energy Transition in the Caribbean: Reshaping the Island Energy Metabolism." *Energy Research and Social Science* 36: 165–74.

———. 2018. "Reassembling Caribbean Energy? Petrocaribe, (Post-)Plantation Sovereignty, and Caribbean Energy Futures." *Journal of Latin American Geography* 17(3): 212–34.

Harvey, David. 1982. *The Limits to Capital*. London: Verso.

Harvey, Penny. 2019. "Infrastructures in and out of Time: The Promise of Roads in Contemporary Peru." In *The Promise of Infrastructure*, edited by Nikhil Anand, Akhil Gupta, and Hannah Appel, 80–101. Durham, NC: Duke University Press.

Hausman, William J., and John L. Neufeld. 1997. "The Rise and Fall of The American and Foreign Power Company: A Lesson from the Past?" *The Electricity Journal* 10(1): 46–53.

Healy, Hali, Joan Martinez-Alier, and Giorgos Kallis. 2015. "From Ecological Modernization to Socially Sustainable Economic Degrowth: Lessons from Ecological Economics." In *International Handbook of Political Ecology*, edited by Raymond L. Bryant, 577–90. Cheltenham, UK: Edward Elgar.

Hearn, Adrian H. 2012. "China, Global Governance and the Future of Cuba." *Journal of Current Chinese Affairs* 41(1): 155–79.

Hecht, Gabrielle. 2009. *The Radiance of France: Nuclear Power and National Identity after World War II*. Cambridge, MA: MIT Press.

———. 2011. "Introduction." In *Entangled Geographies: Empire and Technopolitics in the Global Cold War*, edited by Gabrielle Hecht, 1–12. Cambridge, MA: MIT Press.

———. 2012. *Being Nuclear: Africans and the Global Uranium Trade*. Cambridge, MA: MIT Press.

Hedges, Thomas R. III, Michio Hirano, Katherine Tucker, and Benjamin Caballero. 1997. "Epidemic Optic and Peripheral Neuropathy in Cuba: A Unique Geopolitical Public Health Problem." *Survey of Ophthalmology* 41(4): 341–53.

Heffron, Raphael J., and Darren McCauley. 2017. "The Concept of Energy Justice across the Disciplines." *Energy Policy* 105: 658–67.

Heilbroner, Robert L. 1994. "Do Machines Make History?" In *Does Technology Drive History? The Dilemma of Technological Determinism*, edited by Merritt Roe Smith and Leo Marx, 53–66. Cambridge, MA: MIT Press.

Hernández-Reguant, Ariana. 2009. "Writing the Special Period: An Introduction." In *Cuba in the Special Period: Culture and Ideology in the 1990s*, edited by Ariana Hernández-Reguant, 1–20. New York: Palgrave Macmillan.

———. 2012. "Inventor, Machine, and New Man." In *Caviar and Rum: Cuba-USSR and the Post-Soviet Experience*, edited by Jacqueline Loss and José Manuel Prieto, 199–210. New York: Palgrave Macmillan.

Hickel, Jason. 2020. "The Sustainable Development Index: Measuring the Ecological Efficiency of Human Development in the Anthropocene." *Ecological Economics* 167: 106331.

Hickel, Jason, and Giorgos Kallis. 2020. "Is Green Growth Possible?" *New Political Economy* 25(4): 469–86.

Hirsh, Richard F. 1999. *Power Loss: The Origins of Deregulation and Restructuring in the American Electric Utility System*. Cambridge, MA: MIT Press.

Högselius, Per. 2013. *Red Gas: Russia and the Origins of European Energy Dependence*. New York: Palgrave Macmillan.

Hornborg, Alf. 2001. *The Power of the Machine: Global Inequalities of Economy, Technology, and Environment*. Walnut Creek, CA: AltaMira.

———. 2005. "Undermining Modernity: Protecting Landscapes and Meanings among the Mi'kmaq of Nova Scotia." In *Political Ecology across Spaces, Scales, and Social Groups*, edited by Susan Paulson and Lisa L. Gezon, 196–214. New Brunswick, NJ: Rutgers University Press.

———. 2009. "Zero-Sum World: Challenges in Conceptualizing Environmental Load Displacement and Ecologically Unequal Exchange in the World-System." *International Journal of Comparative Sociology* 50(3–4): 237–62.

———. 2016. *Global Magic: Technologies of Appropriation from Ancient Rome to Wall Street*. Basingstoke, UK: Palgrave Macmillan.

———. 2019. *Nature, Society, and Justice in the Anthropocene: Unraveling the Money-Energy-Technology Complex*. Cambridge: Cambridge University Press.

Hornborg, Alf, Gustav Cederlöf, and Andreas Roos. 2019. "Has Cuba Exposed the Myth of 'Free' Solar Power? Energy, Space, and Justice." *Environment and Planning E: Nature and Space* 2(4): 989–1008.

Howell, Jordan P. 2011. "Powering 'Progress': Regulation and the Development of Michigan's Electricity Landscape." *Annals of the Association of American Geographers* 101(4): 962–70.

Huber, Matthew T. 2013. *Lifeblood: Oil, Freedom, and the Forces of Capital*. Minneapolis: University of Minnesota Press.

———. 2017. "Hidden Abodes: Industrializing Political Ecology." *Annals of the American Association of Geographers* 107(1): 151–66.

———. 2018. "Fossilized Liberation: Energy, Freedom, and the 'Development of the Productive Forces.'" In *Materialism and the Critique of Energy*, edited by Brent Ryan Bellamy and Jeff Diamanti, 501–22. Chicago: MCM'.

Huber, Matthew T., and James McCarthy. 2017. "Beyond the Subterranean Energy Regime? Fuel, Land Use and the Production of Space." *Transactions of the Institute of British Geographers* 42(4): 655–68.

Hughes, Thomas P. 1983. *Networks of Power: Electrification in Western Society, 1880–1930.* Baltimore, MD: Johns Hopkins University Press.

Ingold, Tim. 2018. "From Science to Art and Back Again: The Pendulum of an Anthropologist." *Interdisciplinary Science Reviews* 43(3–4): 213–27.

Irizarry Mora, Edwin. 2015. *Fuentes energéticas: Luchas comunitarias y medioambiente en Puerto Rico.* San Juan: Editorial Universidad de Puerto Rico.

Jasanoff, Sheila, ed. 2006. *States of Knowledge: The Co-Production of Science and Social Order.* London: Routledge.

Jenkins, Kirsten, Darren McCauley, Raphael Heffron, Hannes Stephan, and Robert Rehner. 2016. "Energy Justice: A Conceptual Review." *Energy Research and Social Science* 11: 174–82.

Johnson, Sherry. 2011. *Climate and Catastrophe in Cuba and the Atlantic World in the Age of Revolution.* Chapel Hill: University of North Carolina Press.

Josée Davidson, Mélanie, and Catherine Krull. 2011. "Adapting to Cuba's Shifting Food Landscapes: Women's Strategies of Resistance." *Cuban Studies* 42: 59–77.

Josephson, Paul R. 2002. *Industrialized Nature: Brute Force Technology and the Transformation of the Natural World.* Washington, DC: Island Press.

Kale, Sunila. 2014. *Electrifying India: Regional Political Economies of Development.* Stanford, CA: Stanford University Press.

Kallis, Giorgos. 2011. "In Defence of Degrowth." *Ecological Economics* 70(5): 873–80.

———. 2019. *Limits: Why Malthus Was Wrong and Why the Environmentalists Should Care.* Stanford, CA: Stanford University Press.

Kallis, Giorgos, Vasilis Kostakis, Steffen Lange, Barbara Muraca, Susan Paulson, and Matthias Schmelzer. 2018. "Research on Degrowth." *Annual Review of Environment and Resources* 43: 291–316.

Kapcia, Antoni. 2000. *Cuba: Island of Dreams.* Oxford: Berg.

———. 2005. "Educational Revolution and Revolutionary Morality in Cuba: The 'New Man', Youth and the New 'Battle of Ideas.'" *Journal of Moral Education* 34(4): 399–412.

———. 2008. *Cuba in Revolution: A History since the Fifties.* London: Reaktion Books.

Kingsbury, Donald V. 2016. "Oil's Colonial Residues: Geopolitics, Identity, and Resistance in Venezuela." *Bulletin of Latin American Research* 35(4): 423–36.

———. 2020. "Combined and Uneven Energy Transitions: Reactive Decarbonization in Cuba and Venezuela." *Journal of Political Ecology* 27: 558–79.

Kirk, John M. 2011. "Cuban Medical Cooperation within ALBA: The Case of Venezuela." *International Journal of Cuban Studies* 3(2–3): 221–34.

Klingensmith, Daniel. 2007. *"One Valley and a Thousand": Dams, Nationalism, and Development*. New Delhi: Oxford University Press.

Klinghoffer, Arthur Jay. 1977. *The Soviet Union and International Oil Politics*. New York: Columbia University Press.

Knuth, Sarah. 2019. "Cities and Planetary Repair: The Problem with Climate Retrofitting." *Environment and Planning A: Economy and Space* 51(2): 487–504.

Knuth, Sarah, John Stehlin, and Nate Millington. 2020. "Rethinking Climate Futures through Urban Fabrics: (De)growth, Densification, and the Politics of Scale." *Urban Geography* 41(10): 1335–43.

Kondepudi, Dilip K., and Ilya Prigogine. 1998. *Modern Thermodynamics: From Heat Engines to Dissipative Structures*. Chichester, UK: John Wiley.

Koselleck, Reinhardt. 2004. *Futures Past: On the Semantics of Historical Time*. Translated by Keith Tribe. New York: Columbia University Press.

Lambe, Jennifer L., and Michael J. Bustamante. 2019. "Cuba's Revolution from Within: The Politics of Historical Paradigms." In *The Revolution from Within: Cuba, 1959–1980*, edited by Michael J. Bustamante and Jennifer L. Lambe, 3–32. Durham, NC: Duke University Press.

Larkin, Brian. 2013. "The Politics and Poetics of Infrastructure." *Annual Review of Anthropology* 42: 327–43.

———. 2018. "Promising Forms: The Political Aesthetics of Infrastructure." In *The Promise of Infrastructure*, edited by Nikhil Anand, Akhil Gupta, and Hannah Appel, 176–202. Durham, NC: Duke University Press.

Latouche, Serge. 2009. *Farewell to Growth*. Translated by David Macey. Cambridge: Polity Press.

Lavigne, Marie. 1991. *International Political Economy and Socialism*. Translated by David Lambert. Cambridge: Cambridge University Press.

Lennon, Myles. 2017. "Decolonizing Energy: Black Lives Matter and Technoscientific Expertise amid Solar Transitions." *Energy Research and Social Science* 30: 18–27.

LeoGrande, William M. 2018. "Reversing the Irreversible: President Donald J. Trump's Cuba Policy." *IdeAs: Idées d'Amériques* 10: 1–20.

Levins, Richard. 2008. *Talking about Trees: Science, Ecology and Agriculture in Cuba*. New Delhi: Left Word.

Lewis, Oscar, Ruth M. Lewis, and Susan M. Rigdon. 1977. *Four Women: Living the Revolution—An Oral History of Contemporary Cuba*. Urbana: University of Illinois Press.

Lundgren, Silje. 2011. "Heterosexual Havana: Ideals and Hierarchies of Gender and Sexuality in Contemporary Cuba." PhD diss., Uppsala University, Sweden.

MacEwan, Arthur. 1981. *Revolution and Economic Development in Cuba*. London: Macmillan.

Malm, Andreas. 2016. *Fossil Capital: The Rise of Steam Power and the Roots of Global Warming*. London: Verso.

Malm, Thomas. 2006. "No Island Is an 'Island': Some Perspectives on Human Ecology and Development in Oceania." In *The World System and the Earth System: Global Socioenvironmental Change and Sustainability since the Neolithic*, edited by Alf Hornborg and Carole Crumley, 268–79. Walnut Creek, CA: Left Coast Press.

Mann, Michael. 1984. "The Autonomous Power of the State: Its Origins, Mechanisms and Results." *European Journal of Sociology* 25(2): 185–213.

Maring, Prudensius. 2020. "The Strategy of Shifting Cultivators in West Kalimantan in Adapting to the Market Economy: Empirical Evidence behind Gaps in Interdisciplinary Communication." *Journal of Political Ecology* 27: 1015–35.

Marx, Karl. 1955 (1847). *The Poverty of Philosophy: Answer to "The Philosophy of Poverty" by M. Proudhon*. Translated by the Institute of Marxism-Leninism. Moscow: Progress Publishers.

———. 1973 (1939). *Grundrisse: Foundations of the Critique of Political Economy (Rough Draft)*. Translated by Martin Nicolaus. Harmondsworth, UK: Penguin in association with New Left Review.

———. 1976 (1867). *Capital: A Critique of Political Economy* 1. Translated by Ben Fowkes. Harmondsworth, UK: Penguin in association with New Left Review.

Masera, Omar, Barbara Saatkamp Taylor, and Daniel M. Kammen. 2000. "From Linear Fuel Switching to Multiple Cooking Strategies: A Critique and an Alternative to the Energy Ladder Model." *World Development* 28(12): 2083–103.

Massey, Doreen. 1993. "Power-Geometry and a Progressive Sense of Place." In *Mapping the Futures: Local Cultures, Global Change*, edited by John Bird, Barry Curtis, Tim Putnam, and Lisa Tickner, 59–69. London: Routledge.

———. 2005. *For Space*. London: Sage.

———. 2007. *World City*. Cambridge: Polity Press.

Maurer, Noel. 2013. *The Empire Trap: The Rise and Fall of U.S. Intervention to Protect American Property Overseas, 1893–2013*. Princeton, NJ: Princeton University Press.

McGillivray, Gillian. 2009. *Blazing Cane: Sugar Communities, Class, and State Formation in Cuba, 1868–1959*. Durham, NC: Duke University Press.

McNeill, J. R. 2010. "The First Hundred Thousand Years." In *The Turning Points of Environmental History*, edited by Frank Uekotter, 13–28. Pittsburgh, PA: University of Pittsburgh Press.

McNeill, J. R., and Peter Engelke. 2016. *The Great Acceleration: An Environmental History of the Anthropocene since 1945*. Cambridge, MA: Belknap Press.

Meadows, Donella J., Dennis L. Meadows, Jørgen Randers, and William W. Behrens III. 1972. *The Limits to Growth: A Report for The Club of Rome's Project on the Predicament of Mankind*. New York: Universe.

Meehan, Katie M. 2014. "Tool-Power: Water Infrastructure as Wellsprings of State Power." *Geoforum* 57: 215–24.

Mehrotra, Santosh. 1990. *India and the Soviet Union: Trade and Technology Transfer*. Cambridge: Cambridge University Press.

Mesa-Lago, Carmelo. 1998. "Assessing Economic and Social Performance in the Cuban Transition of the 1990s." *World Development* 26(5): 857–76.

Miller, Ian Jared, and Paul Warde. 2019. "Energy Transitions as Environmental Events." *Environmental History* 24(3): 464–71.

Mitchell, Timothy. 1999. "Society, Economy, and the State Effect." In *State/ Culture: State-Formation after the Cultural Turn*, edited by George Steinmetz, 76–97. Ithaca, NY: Cornell University Press.

———. 2002. *Rule of Experts: Egypt, Techno-Politics, Modernity*. Berkeley: University of California Press.

———. 2011. *Carbon Democracy: Political Power in the Age of Oil*. London: Verso.

Mommer, Bernard. 2003. "Subversive Oil." In *Venezuelan Politics in the Chávez Era: Class, Polarization, and Conflict*, edited by Steve Ellner and Daniel Hellinger, 131–46. Boulder, CO: Lynne Rienner Publishers.

Moreno Fraginals, Manuel. 1976. *The Sugarmill: The Socioeconomic Complex of Sugar in Cuba 1760–1860*. Translated by Cedric Belfrage. New York: Monthly Review Press.

Morley, Morris H. 1987. *Imperial State and Revolution: The United States and Cuba, 1952–1986*. Cambridge: Cambridge University Press.

Murphy, Catherine. 1999. *Cultivating Havana: Urban Agriculture and Food Security in the Years of Crisis*. Development Report 12. Oakland, CA: Food First.

Newell, Peter. 2019. "*Trasformismo* or Transformation? The Global Political Economy of Energy Transitions." *Review of International Political Economy* 26(1): 25–48.

———. 2021. "Race and the Politics of Energy Transitions." *Energy Research and Social Science* 71: 1–5.

Nye, David E. 1990. *Electrifying America: Social Meanings of a New Technology, 1880–1940*. Cambridge, MA: MIT Press.

———. 1998. *Consuming Power: A Social History of American Energies*. Cambridge, MA: MIT Press.

O'Brien, Thomas F. 1993. "The Revolutionary Mission: American Enterprise in Cuba." *The American Historical Review* 98(3): 765–85.

O'Neill, Daniel W., Andrew L. Fanning, William F. Lamb, and Julia K. Steinberger. 2018. "A Good Life for All within Planetary Boundaries." *Nature Sustainability* 1: 88–95.

Overland, Indra. 2019. "The Geopolitics of Renewable Energy: Debunking Four Emerging Myths." *Energy Research and Social Science* 49: 36–40.

Padrón Hernández, Maria. 2012. "Beans and Roses: Everyday Economies and Morality in Contemporary Havana, Cuba." PhD diss., University of Gothenburg, Sweden.

Painter, Joe. 2006. "Prosaic Geographies of Stateness." *Political Geography* 25(7): 752–74.

———. 2010. "Rethinking Territory." *Antipode* 42(5): 1090–118.

Parenti, Christian. 2015. "The Environment Making State: Territory, Nature, and Value." *Antipode* 47(4): 829–48.

Paulson, Susan. 2017. "Degrowth: Culture, Power and Change." *Journal of Political Ecology* 24: 425–48.

Peet, Richard, and Elaine Hartwick. 2009. *Theories of Development: Contentions, Arguments, Alternatives*, 2nd ed. New York: The Guilford Press.

Pérez, Louis A. Jr. 1999. *On Becoming Cuban: Identity, Nationality and Culture.* New York: HarperCollins.

———. 2001. *Winds of Change: Hurricanes and the Transformation of Nineteenth-Century Cuba.* Chapel Hill: University of North Carolina Press.

———. 2011. *Cuba: Between Reform and Revolution*, 4th ed. Oxford: Oxford University Press.

Pérez-López, Jorge F. 1987. "Cuban Oil Reexports: Significance and Prospects." *The Energy Journal* 8(1): 1–16.

———. 1991. *The Economics of Cuban Sugar.* Pittsburgh, PA: University of Pittsburgh Press.

———. 1993. "Cuba's Thrust to Attract Foreign Investment: A Special Labor Regime for Joint Ventures in International Tourism." *University of Miami Inter-American Law Review* 24(2): 221–79.

Pérez-Stable, Marifeli. 2012. *The Cuban Revolution: Origins, Course, and Legacy*, 3rd ed. Oxford: Oxford University Press.

Peters, F. W. 1920. "Electrification of the Hershey Cuban Railway: A High Voltage Direct Current System." *Scientific American Monthly*, June: 540–42.

Pfeiffer, Dale Allen. 2006. *Eating Fossil Fuels: Oil, Food and the Coming Crisis in Agriculture.* Gabriola Island, BC: New Society Publishers.

Phalkey, Jahnavi. 2013. *Atomic State: Big Science in Twentieth-Century India.* Ranikhet, India: Permanent Black.

Piercy, Emma, Rachel Granger, and Chris Goodier. 2010. "Planning for Peak Oil: Learning from Cuba's 'Special Period.'" *Urban Design and Planning* 163(4): 169–76.

Pollitt, Brian H., and G. B. Hagelberg. 1994. "The Cuban Sugar Economy in the Soviet Era and after." *Cambridge Journal of Economics* 18(6): 547–69.

Pomeranz, Kenneth. 2000. *The Great Divergence: China, Europe, and the Making of the Modern World Economy.* Princeton, NJ: Princeton University Press.

Powell, Kathy. 2008. "Neoliberalism, the Special Period and Solidarity in Cuba." *Critique of Anthropology* 28(2): 177–97.

Power, Marcus, and Joshua Kirshner. 2019. "Powering the State: The Political Geographies of Electrification in Mozambique." *Environment and Planning C: Politics and Space* 37(3): 498–518.

Premat, Adriana. 2012. *Sowing Change: The Making of Havana's Urban Agriculture.* Nashville, TN: Vanderbilt University Press.

Princen, Thomas, Jack P. Manno, and Pamela L. Martin, eds. 2015. *Ending the Fossil Fuel Era.* Cambridge, MA: MIT Press.

Rabinbach, Anson. 1992. *The Human Motor: Energy, Fatigue, and the Origins of Modernity.* Berkeley: University of California Press.

Ramos Guadalupe, Luis E. 2011. *Fidel Castro ante los desastres naturales: Pensamiento y acción.* Havana: Oficina de Publicaciones del Consejo de Estado.

Ratliff, William. 1990. "Cuban Foreign Policy toward Far East and Southeast Asia." In *Cuba: The International Dimension*, edited by Georges Fauriol and Eva Loser, 205–32. New Brunswick, NJ: Transaction Publishers.

Ratnieks, Henry. 1976. "Baltic Oil Prospects and Problems." *Journal of Baltic Studies* 7(4): 312–19.

Reisinger, William M. 1992. *Energy and the Soviet Bloc: Alliance Politics after Stalin.* Ithaca, NY: Cornell University Press.

Ríos, Arcadio, and Félix Ponce. 2002. "Mechanization, Animal Traction, and Sustainable Agriculture." In *Sustainable Agriculture and Resistance: Transforming Food Production in Cuba*, edited by Fernando Funes, Luis García, Martin Bourque, Nilda Pérez, and Peter Rosset, 155–63. Oakland, CA: Food First Books.

Robbins, Paul. 2008. "The State in Political Ecology: A Postcard to Political Geography from the Field." In *The SAGE Handbook of Political Geography*, edited by Kevin R. Cox, Murray Low, and Jennifer Robinson, 205–18. London: Sage.

———. 2020. "Is Less More . . . or Is More Less? Scaling the Political Ecologies of the Future." *Political Geography* 76: 102018.

Rodríguez, Eduardo Luis. 2000. *The Havana Guide: Modern Architecture 1925–1965.* Translated by Lona Scott Fox. New York: Princeton Architectural Press.

Rodríguez Castellón, Santiago. 1988. "La industria eléctrica cubana." *Economía y Desarrollo* 18(5): 150–59.

Rojas, Rafael. 2019. "The New Text of the Revolution." In *The Revolution from Within: Cuba, 1959–1980*, edited by Michael J. Bustamante and Jennifer L. Lambe, 33–46. Durham, NC: Duke University Press.

Roman, Peter. 2003. *People's Power: Cuba's Experience with Representative Government, Updated Edition.* Lanham, MD: Rowman and Littlefield.

Rosendahl, Mona. 1997. *Inside the Revolution: Everyday Life in Socialist Cuba.* Ithaca, NY: Cornell University Press.

Rosset, Peter M., and Valentín Val. 2019. "The 'Campesino a Campesino' Agroecology Movement in Cuba: Food Sovereignty and Food as a Commons." In *Routledge Handbook of Food as a Commons*, edited by Jose Luis Vivero-Pol, Tomaso Ferrando, Olivier de Schutter, and Ugo Mattei, 251–65. London: Routledge.

Rosset, Peter Michael, Braulio Machín-Sosa, Adilén María Roque Jaime, and Dana Rocío Ávila Lozano. 2011. "The Campesino-to-Campesino Agroecology Movement of ANAP in Cuba: Social Process Methodology in the Construction of Sustainable Peasant Agriculture and Food Sovereignty." *Journal of Peasant Studies* 38(1): 161–91.

Rupprecht, Tobias. 2011. "Socialist High Modernity and Global Stagnation: A Shared History of Brazil and the Soviet Union during the Cold War." *Journal of Global History* 6(3): 505–28.

Saney, Isaac. 2004. *Cuba: A Revolution in Motion*. London: Zed.

Scheer, Hermann. 2002. *The Solar Economy: Renewable Energy for a Sustainable Global Future*. Translated by Andrew Ketley. London: Earthscan.

———. 2007. *Energy Autonomy: The Economic, Social and Technological Case for Renewable Energy*. Translated by Jeremiah M. Riemer. London: Earthscan.

Scheidel, Arnim, and Alevgul H. Sorman. 2012. "Energy Transitions and the Global Land Rush: Ultimate Drivers and Persistent Consequences." *Global Environmental Change* 22: 588–95.

Schmid, Sonja D. 2011. "Nuclear Colonization? Soviet Technopolitics in the Second World." In *Entangled Geographies: Empire and Technopolitics in the Global Cold War*, edited by Gabrielle Hecht, 125–54. Cambridge, MA: MIT Press.

———. 2015. *Producing Power: The Pre-Chernobyl History of the Soviet Nuclear Industry*. Cambridge, MA: MIT Press.

Schroeder, Gertrude E. 1990. "Economic Reform of Socialism: The Soviet Record." *Annals of the American Academy of Political and Social Science* 507(1): 35–43.

Schwartzman, David. 1996. "Solar Communism." *Science and Society* 60(3): 307–31.

Schwenkel, Christina. 2018. "The Current Never Stops: Intimacies of Energy Infrastructure in Vietnam." In *The Promise of Infrastructure*, edited by Nikhil Anand, Akhil Gupta, and Hannah Appel, 102–29. Durham, NC: Duke University Press.

Scott, James C. 1998. *Seeing Like a State: How Certain Schemes to Improve the Human Condition Have Failed*. New Haven, CT: Yale University Press.

———. 2006. "High Modernist Social Engineering: The Case of the Tennessee Valley Authority." In *Experiencing the State*, edited by Lloyd I. Rudolph and John Kurt Jacobsen, 3–52. New Delhi: Oxford University Press.

Shearman, Peter. 1989. "Gorbachev and the Restructuring of Soviet-Cuban Relations." *Journal of Communist Studies* 5(4): 63–83.

Shirley, Rebekah, and Daniel Kammen. 2013. "Renewable Energy Sector Development in the Caribbean: Current Trends and Lessons from History." *Energy Policy* 57: 244–52.

Shove, Elizabeth, and Gordon Walker. 2014. "What Is Energy For? Social Practice and Energy Demand." *Theory, Culture and Society* 31(5): 41–58.

Siebert, Fred S., Theodore Peterson, and Wilbur Schramm. 1963. *Four Theories of the Press: The Authoritarian, Libertarian, Social Responsibility, and Soviet Communist Concepts of What the Press Should Be and Do.* Urbana: University of Illinois Press.

Sieferle, Rolf Peter. 2001. *The Subterranean Forest: Energy Systems and the Industrial Revolution.* Translated by Michael P. Osman. Cambridge: White Horse Press.

Silver, Jonathan. 2015. "Disrupted Infrastructures: An Urban Political Ecology of Interrupted Electricity Access in Accra." *International Journal of Urban and Regional Research* 39(5): 984–1003.

Silverman, Bertram, ed. 1971. *Man and Socialism in Cuba: The Great Debate.* New York: Atheneum.

Simone, AbdouMaliq. 2012. "Infrastructure: Commentary by AbdouMaliq Simone." *Cultural Anthropology*, Curated Collection: Infrastructure, edited by Jessica Lockrem and Adonia Lugo. https://journal.culanth.org/index.php /ca/infrastructure-abdoumaliq-simone.

Smil, Vaclav. 2015. *Power Density: A Key to Understanding Energy Sources and Uses.* Cambridge, MA: MIT Press.

———. 2019. *Growth: From Microorganisms to Megacities.* Cambridge, MA: MIT Press.

Smith-Nonini, Sandy. 2020. "The Debt/Energy Nexus behind Puerto Rico's Long Blackout." *Latin American Perspectives* 232(47): 64–86.

Sovacool, Benjamin K. 2014. "What Are We Doing Here? Analyzing Fifteen Years of Energy Scholarship and Proposing a Social Science Research Agenda." *Energy Research and Social Science* 1: 1–29.

Sovacool, Benjamin K., and Michael H. Dworkin. 2015. "Energy Justice: Conceptual Insights and Practical Applications." *Applied Energy* 142: 435–44.

Sparks, Colin. 1998. *Communism, Capitalism and the Mass Media.* London: Sage.

Star, Susan Leigh. 1999. "The Ethnography of Infrastructure." *American Behavioral Scientist* 43(3): 377–91.

Stengers, Isabelle. 2010. *Cosmopolitics I.* Minneapolis: University of Minnesota Press.

Stites, Richard. 1989. *Revolutionary Dreams: Utopian Vision and Experimental Life in the Russian Revolution.* Oxford: Oxford University Press.

Stricker, Pamela. 2007. *Toward a Culture of Nature: Environmental Policy and Sustainable Development in Cuba.* Lanham, MD: Lexington Books.

Susman, Paul. 1987. "Spatial Equality in Cuba." *International Journal of Urban and Regional Research* 11(2): 218–42.

Sutela, Pekka. 1991. *Economic Thought and Economic Reform in the Soviet Union*. Cambridge: Cambridge University Press.

Swyngedouw, Erik. 2015. *Liquid Power: Contested Hydro-Modernities in Twentieth-Century Spain*. Cambridge, MA: MIT Press.

Thoyre, Autumn. 2021. "Negawatt Resource Frontiers: Extracting Energy Efficiency from Private Spaces." *Environment and Planning E: Nature and Space* 4(4): 1703–23.

Torres Pérez, Ricardo. 2017. "Updating the Cuban Economy: The First 10 Years." *Social Research: An International Quarterly* 84(2): 255–75.

Toye, J. F. J., and Richard Toye. 2003. "The Origins and Interpretation of the Prebisch-Singer Thesis." *History of Political Economy* 35(3): 437–67.

Travieso Pedroso, Daniel, and Marin Kaltschmitt. 2012. "*Dichrostachys cinerea* as Possible Energy Crop—Facts and Figures." *Biomass Conversion and Biorefinery* 2(1): 41–51.

Urry, John. 2013. *Societies beyond Oil: Oil Dregs and Social Futures*. London: Zed.

Vakulchuk, Roman, Indra Overland, and Daniel Scholten. 2020. "Renewable Energy and Geopolitics: A Review." *Renewable and Sustainable Energy Reviews* 122: 1–12.

Vayda, Andrew P. 1983. "Progressive Contextualization: Methods for Research in Human Ecology." *Human Ecology* 11(3): 265–81.

von Schnitzler, Antina. 2013. "Traveling Technologies: Infrastructure, Ethical Regimes, and the Materiality of Politics in South Africa." *Cultural Anthropology* 28(4): 670–93.

———. 2018. "Infrastructure, Apartheid Technopolitics, and Temporalities of 'Transition.'" In *The Promise of Infrastructure*, edited by Nikhil Anand, Akhil Gupta, and Hannah Appel, 133–54. Durham, NC: Duke University Press.

Walker, Gordon. 2021. *Energy and Rhythm: Rhythmanalysis for a Low Carbon Future*. Lanham, MD: Rowman and Littlefield.

Wendling, Amy E. 2009. *Karl Marx on Technology and Alienation*. Basingstoke, UK: Palgrave Macmillan.

Whittle, Daniel, and Orlando Rey Santos. 2006. "Protecting Cuba's Environment: Efforts to Design and Implement Effective Environmental Laws and Policies in Cuba." *Cuban Studies* 37: 73–103.

Wilpert, Gregory. 2007. *Changing Venezuela by Taking Power: The History and Policies of the Chávez Government*. London: Verso.

Wilson, Marisa. 2014. *Everyday Moral Economies: Food, Politics and Scale in Cuba*. Chichester, UK: Wiley-Blackwell.

Winner, Langdon. 1986. *The Whale and the Reactor: A Search for Limits in an Age of High Technology*. Chicago: University of Chicago Press.

Winpenny, Thomas R. 1995. "Milton S. Hershey Ventures into Cuban Sugar." *Pennsylvania History* 62(4): 491–502.

Wright, Julia. 2009. *Sustainable Agriculture and Food Security in an Era of Oil Scarcity: Lessons from Cuba.* London: Earthscan.

Wrigley, E. A. 2016. *The Path to Sustained Growth: England's Transition from an Organic Economy to an Industrial Revolution.* Cambridge: Cambridge University Press.

WWF. 2006. *Living Planet Report.* Gland, Switzerland: WWF International.

Yaffe, Helen. 2009. *Che Guevara: The Economics of Revolution.* Basingstoke, UK: Palgrave Macmillan.

———. 2011. "The Bolivarian Alliance for the Americas: An Alternative Development Strategy." *International Journal of Cuban Studies* 3(2–3): 128–44.

———. 2020. *We Are Cuba! How a Revolutionary People Have Survived in a Post-Soviet World.* New Haven, CT: Yale University Press.

Yoss. 2012. "What the Russians Left Behind." Translated by Daniel W. Koon. In *Caviar with Rum: Cuba-USSR and the Post-Soviet Experience,* edited by Jacqueline Loss and José Manuel Prieto, 211–25. New York: Palgrave Macmillan.

Zanetti, Oscar, and Alejandro García. 1998. *Sugar and Railroads: A Cuban History, 1837–1959.* Translated by Franklin W. Knight and Mary Todd. Chapel Hill: University of North Carolina Press.

Zhang, Shuai, and Dajian Zhu. 2022. "Incorporating 'Relative' Ecological Impacts into Human Development Evaluation: Planetary Boundaries-Adjusted HDI." *Ecological Indicators* 137: 108786.

Zimmerer, Karl S. 2000. "Rescaling Irrigation in Latin America: The Cultural Images and Political Ecology of Water Resources." *Ecumene* 7(2): 150–75.

Index

ethics, revolutionary: energy efficiency and, 18, 138, 141; immoral behavior and, 139, 150–52, 157, 163; sacrifice and, 37, 44, 139; thrift and, 18, 58–59, 138, 141, 162

forests, 9, 21–22, 93, 99, 103, 161; deforestation, 10–11, 16, 21–22, 39. *See also* charcoal; wood

gasoline, ix, 99, 147, 149, 179n71; availability of, 3, 68–69; efficiency of use, 60, 106, 144–45
gender, 6, 17, 72–74, 123; energy demand and, 25, 50; energy transition and, 18, 161; inequality in the special period, 15–17, 72–75. *See also* cooking
General Electric, 23, 25
geopolitics: energy transition and, 18, 69, 92, 97, 168; entropy and, 11–13, 168; environmental limits and, 7, 167–69; special period and, 3–5, 7, 69, 71, 79–80, 84–85, 97, 118–19
German Democratic Republic. *See* East Germany
GOELRO, 31–33, 49, 57
Gorbachev, Mikhail, 77–78
Great Debate in Cuba, 34–38, 78
growth, economic, 5–8, 18, 44; abundance and, 7–8, 167; communism, transition to, and, 56, 160, 167–68; energy efficiency and, 136–38, 142, 149. *See also* degrowth
Guevara, Ernesto Che, 27, 34–38, 45–46, 139; economic management and, 36–38, 51; electrification and, 1–2, 29–30, 32–33, 44, 48–49; productive forces and, 32, 36–37, 117
guevarista ideals, 38, 46, 77, 139

hard currency, 80, 82, 87, 154; as state income, 40, 56–58, 66, 77, 80, 96, 137
Harrison, Conor, 24, 169
Harvey, Penny, 15, 52
Havana, 21, 29*fig.*, 42, 45, 70, 122, 156; electrification of, 2, 22, 46*fig.*, 125*fig.*; infrastructure, 96, 197n98; transport, urban, 110, 144
heat loss (transmission), 47, 55, 81, 127–29, 140
Hickel, Jason, 6
hitchhiking, 88*fig. See also* transport
Hornborg, Alf, 12–13, 52, 79, 165
Huber, Matthew, 51, 167, 179n71
hurricane(s), 21; damage, 4, 111, 122–24, 157–58; resilience, 111, 123, 128–29, 149–51, 155

hydroelectric dam, 32, 39–40, 61, 63–64, 98, 107, 127*fig.*, 181n21

incentives, 36, 57–60, 101, 117, 152
India, 13–14, 41, 62, 80
industrialization, 11, 19, 57, 60, 141, 150; electrification and, 2, 30, 32–33, 35, 50; import substitution, 26–29; socialist development and, 6, 27, 35, 81, 162; unequal exchange and, 41–42, 76–77
inequality, 16–17, 71–75, 88, 91, 154; of trade, 41, 132. *See also* gender; race; unequal exchange
infrastructural form, 2, 15–17, 31, 54–55, 65–66, 100, 160; energy use and, 83, 119, 164; fracturing of, 85, 90, 94; political economy of, 106, 114, 170; spatiality of, 12, 48–49, 51, 85, 130
infrastructural power, 13–14, 54, 114, 149, 164, 169
infrastructure: control, 14–15, 26, 114–18, 129–30, 133–34, 149, 166; spatial (de)concentration of, 31–32, 47, 66, 117, 124, 129–30, 165–66. *See also* centralization; decentralization
Ingold, Tim, 13, 52, 114
interconnectivity, 2, 47–49, 55, 117, 165
investment, foreign, 23–24, 26, 39, 96–97, 159
irrigation, 107, 115, 119
island geography, 10–12, 16, 40, 42, 126, 168–70; import dependence and, 17, 39–40, 132; neocolonialism and, 26, 39, 79, 132; territorialization and, 94–95, 106, 126, 158–59, 168–69

joint venture(s), 96, 106, 133–34, 137, 145, 159, 197n85
Juraguá. *See* nuclear energy, Juraguá plant

Kammen, Daniel, 52, 115–16
kerosene, 3, 22, 80, 112; as cooking fuel, 10, 50, 70, 99, 102, 105, 119; as entitlement, 51, 68; rationing of, 3, 51, 70, 123, 149–50
Khrushchev, Nikita, 36, 40
Knuth, Sarah, 137
Koselleck, Reinhart, 85

labor, 6, 14, 18, 26, 33, 36–38, 123; degrowth and, 6, 162–63, 167–68; liberation from, 33–34, 119, 161; Marx and, 30, 32, 160–61. *See also* socialist state
land use, 10–11, 101, 103–4, 120
law of value, 33, 36–37, 42, 161

Founded in 1893,
UNIVERSITY OF CALIFORNIA PRESS
publishes bold, progressive books and journals
on topics in the arts, humanities, social sciences,
and natural sciences—with a focus on social
justice issues—that inspire thought and action
among readers worldwide.

The UC PRESS FOUNDATION
raises funds to uphold the press's vital role
as an independent, nonprofit publisher, and
receives philanthropic support from a wide
range of individuals and institutions—and from
committed readers like you. To learn more, visit
ucpress.edu/supportus.

www.ingramcontent.com/pod-product-compliance
Lightning Source LLC
Chambersburg PA
CBHW020851270326
41928CB00006B/651